BASIC

ELECTRICITY/ELECTRONICS

VOLUME 4

Understanding & Using Test Instruments

By

Training & Retraining, Inc.

Second Edition Revised by

Joseph A. Risse, P.E.

Howard W. Sams & Co., Inc.

4300 WEST 62ND ST. INDIANAPOLIS, INDIANA 46268 USA

SECOND EDITION
FIRST PRINTING—1982

International Standard Book Number: 0-672-21504-7
Library of Congress Catalog Card Number: 80-50045

Edited by: *Frank N. Speights*
Illustrated by: *T. R. Emrick*

Printed in the United States of America.

Acknowledgments

Grateful acknowledgment is made to all those who participated in the preparation, compilation, and editing of this series. Without their valuable contributions this series would not have been possible.

In this regard, prime consideration is due Bernard C. Monnes, Educational Specialist, Navy Electronics School, for his excellent contributions in the areas of writing, editorial organization, and final review of the entire series. The finalization of these volumes, both as to technical content and educational value, is due principally to his tireless and conscientious efforts.

Grateful appreciation is also extended to Lt. Loren Worley, USN, and Ashley G. Skidmore, BUSHIPS, Dept. of the Navy, for their original preparatory contributions and co-editing of this series. We also want to thank Irene and Don Koosis, Raymond Mungiu, George V. Novotny, and Robert J. Brite for their technical writing and contribution to the programmed method of presentation. Special thanks to Robert L. Snyder for his initial preparation and organizational work on the complete series.

Finally, special thanks are due the Publisher's editorial staff for invaluable assistance beyond the normal publisher-author relationship.

TRAINING & RETRAINING, INC.

Contents

CHAPTER 3

CHAPTER 4

CHAPTER 5

CHAPTER 6

CHAPTER 7

CHAPTER 8

Introduction

This fourth volume in the series covers test equipment, a specialized and highly important part of the field of electronics. In addition to explaining how each type of measuring equipment works, the text also discusses the proper way to use many of the instruments. Examples of practical applications are given. In this way, your interest in the subject is maintained, and the learning process is made easier. Your study is centered around the principles and design of commonly used test instruments. Thus, the knowledge you gain from this volume can be put to practical use in troubleshooting electrical and electronic equipment.

WHAT YOU WILL LEARN

The most important types of electronic test equipment are discussed in considerable detail in this volume. You will learn the operating principles of the voltmeter, ammeter, and ohmmeter, and you will see how the functions of these three instruments are combined in the multimeter. The operation of the basic meter movement is explained and a discussion of meter-scale reading is included.

Both vacuum-tube and solid-state electronic voltmeters and multimeters are covered. The fundamentals of the now-popular digital instruments, and methods for using these instruments, will also be described.

You will learn that the oscilloscope is able to display waveforms for study and that, for this reason, it is a widely used test instrument. The ways in which the oscilloscope functions are described in detail. The cathode-ray tube, sawtooth generators, amplifiers, and other important parts of the oscilloscope are all discussed. The text explains the uses of the oscilloscope in making measurements and in troubleshooting.

Tube and transistor testers are also covered. You will learn how they work, and you will be shown their capabilities and their limitations.

One chapter is devoted to bridge instruments. The discussion will show how they can be used to measure resistance, capacitance, inductance, and impedance.

You will learn how signal generators provide a controllable signal source for use in troubleshooting and maintaining electronic equipment. Audio-frequency generators, function generators, and radio-frequency signal generators are discussed, and the text shows how each type can be effectively used. The proper way to align a radio receiver is given. The signal-substitution and signal-tracing methods of troubleshooting are explained.

One of the most important uses of electronic test equipment is in troubleshooting other electronic equipment, including other test instruments. The final chapter of this volume is devoted to the subject of logical troubleshooting. You will learn how you can use test equipment along with your knowledge of electronic principles to conduct a systematic search for a defect in a piece of electronic equipment.

WHAT YOU SHOULD KNOW BEFORE YOU START

Before you start to study this book, it is essential that you have a good background in the principles of electricity and electronics, including the fundamentals of tube and transistor circuits. This background can be obtained by studying the first three volumes of this series. With the proper background, you should have no trouble understanding this text. All new terms are carefully defined. Enough math is used to give a precise interpretation of important principles, but if you know how to add, subtract, multiply, and divide, the mathematical expressions will give you no trouble.

WHY THE TEXT FORMAT WAS CHOSEN

During the past few years, new concepts of learning have been developed under the common heading of programmed instruction. Although there are arguments for and against each of the several formats or styles of programmed textbooks, the value of programmed instruction itself has been proved to be sound. Most educators now seem to agree that a style of programming should be developed to fit the needs for teaching a particular subject. To help you progress successfully through this volume, a brief explanation of the programmed format follows.

Each chapter is divided into small bits of information presented in a sequence that has been proved best for learning purposes. Some of the information bits are very short—a single sentence in some cases. Others may include several paragraphs. The length of each presentation is determined by the nature of the concept being explained and the knowledge the reader has gained up to that point.

The text is designed around two-page segments. Facing pages include information on one or more concepts, complete with illustrations designed to clarify the word descriptions used. Self-testing questions are included in most of these two-page segments. Many of these questions are in the form of statements that require that you fill in one or more missing words; other questions are either multiple-choice or simple essay types. Answers are given on the succeeding page, so you will have the opportunity to check the accuracy of your response and verify what you have or have not learned before proceeding. When you find that your answer to a question does not agree with that given, you should restudy the information to determine why your answer was incorrect. As you can see, this method of question–answer programming ensures that you will advance through the text as quickly as you are able to absorb what has been presented.

The beginning of each chapter features a preview of its contents, and a review of the important points is contained at the end of the chapter. The preview gives you an idea of the purpose of the chapter—what you can expect to learn. This helps to give a practical meaning to the information as it is presented. The review at the completion of the chapter summarizes its content so that you can locate and restudy those areas which have escaped your full comprehension. And, just as important, the review is a definite aid to the retention and recall of what you have learned.

HOW YOU SHOULD STUDY THIS TEXT

Naturally, good study habits are important. You should set aside a specific time each day to study in an area where you can concentrate without being disturbed. Select a time when you are at your mental peak, a period when you feel most alert.

Here are a few pointers you will find helpful in getting the most out of this volume.

1. Read each sentence carefully and deliberately. There are no unnecessary words or phrases; each sentence presents or supports a thought that is important to your understanding of electricity and electronics.
2. When you are referred to or come to an illustration, stop at the end of the sentence you are reading and study the illustration. Make sure you have a mental picture of its general content. Then continue reading, returning to the illustration each time a detailed examination is required. The drawings were especially planned to reinforce your understanding of the subject.
3. At the bottom of most right-hand pages, you will find one or more questions to be answered. Some of these contain "fill-in" blanks. Since more than one word might logically fill a given blank, the number of dashes indicates the number of letters in the desired word. In answering the questions, it is important that you actually do so in writing, either in the book or on a separate sheet of paper. The physical act of writing the answers provides greater retention than merely thinking the answer. Writing will not become a chore since most of the required answers are short.
4. Answer all questions in a section before turning the page to check the accuracy of your responses. Refer to any of the material you have read if you need help. If you don't know the answer even after a quick review of the related text, finish answering any remaining questions. Then, if

the answers to any questions that you have skipped still haven't come to you, turn the page and check the answer section.

5. When you have answered a question incorrectly, return to the appropriate paragraph or page and restudy the material. Knowing the correct answer to a question is less important than understanding **why** it is correct. Each section of new material is based on previously presented information. If there is a weak link in this chain, the later material will be more difficult to understand.
6. In some instances, the text describes certain principles in terms of the results of simple experiments. The information is presented so that you will gain knowledge whether you perform the experiments or not. However, you will gain a greater understanding of the subject if you do perform the suggested experiments.
7. Carefully study the review, "What You Have Learned," at the end of each chapter. This review will help you gauge your knowledge of the information in the chapter and actually reinforce your knowledge. When you run across statements that you don't completely understand, reread the sections relating to these statements, and recheck the questions and answers before going to the next chapter.

This volume has been carefully planned to make the learning process as easy as possible. Naturally, a certain amount of effort on your part is required if you are to obtain the maximum benefit from the book. However, if you follow the pointers just given, your efforts will be well rewarded, and you will find that your study of electricity and electronics will be a pleasant and interesting experience.

1

Multimeters

what you will learn

In this chapter, you will learn the functions and operations of the multimeter. You will study the analog multimeter, in particular, as that is the type meant when the word "multimeter" is mentioned. Later, we will discuss the digital multimeter. Briefly, with an analog meter, the value being measured is interpreted by a pointer moving across a calibrated scale. In a digital meter, the exact value is shown as a numerical display. The multimeter is the instrument most frequently used by electronic technicians. After finishing this chapter, you will be able to measure voltage, current, resistance, and (with some multimeters) other electrical characteristics. You will learn how to use these instruments for their intended purpose, the transfer of information from the circuit to the technician.

TEST EQUIPMENT MOST OFTEN USED

There are several basic classes of test equipment. The multimeter is an example of one class; oscilloscopes represent another. There are many types, or models, in each class. Some types have a greater precision than others. Some instruments are easier to set up or use than others. Equipment may also vary in the type or amount of information that the instrument can provide.

Types of instruments within a test-equipment class may vary, but the way they function and the procedures for using them correctly are basically the same. For this reason, there is no need to learn the step-by-step procedures for using each of the hundreds of different models. A technician can gain ability in electronics by first learning a set of underlying fundamentals and, then, developing skill through practice in applying them.

BASIC CONCEPTS

A **multimeter** combines the features of a **voltmeter**, an **ammeter**, and an **ohmmeter** in a single instrument having but one meter movement. See Fig. 1-1.

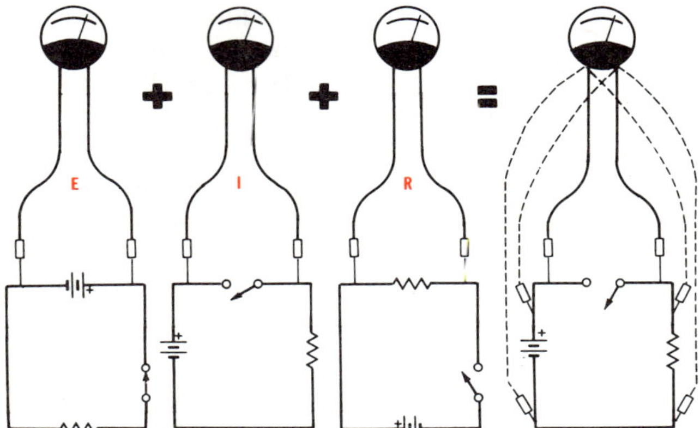

Fig. 1-1. A multimeter is several meters in one.

A multimeter, or multitester, can be used to measure voltage, current, and resistance within the limits of several ranges of values. From the technician's point of view, all analog multimeters consist of three basic sections—meter, circuitry, and front panel. The meter coil moves a pointer across a calibrated scale to a mark that indicates the measurement value. The circuitry is a network of components that determines the functions (ohmmeter, ammeter, voltmeter) and ranges. The front panel contains the controls and jacks that permit operation of the instrument (see Fig. 1-2).

Most analog meters have moving-coil movements. As the name implies, the movement has a coil of wire that is free to rotate between the north- and south-seeking poles of a permanent magnet. Current flowing through the coil sets up a magnetic field. This field reacts with the field existing between the

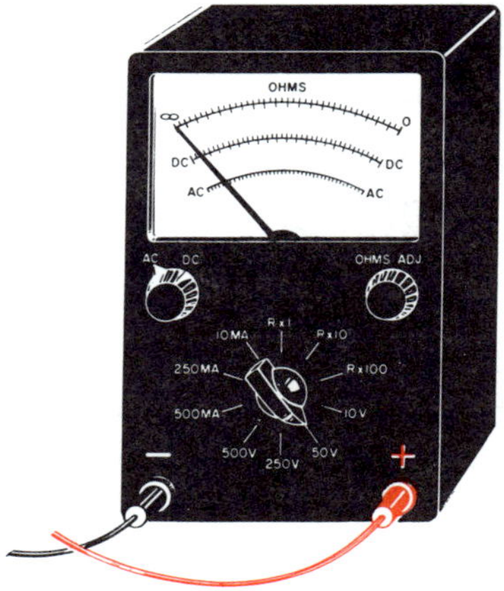

Fig. 1-2. A multimeter.

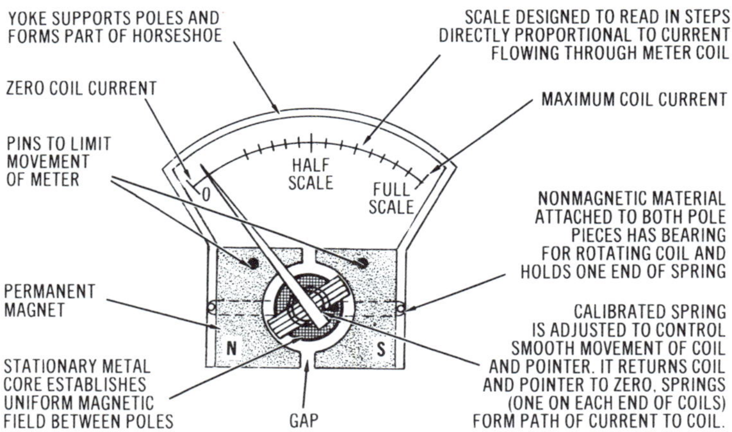

Fig. 1-3. Basic construction of a meter.

poles of the magnet and causes the coil to rotate. A pointer attached to the coil moves to a position on the meter scale; the position of the pointer depends on the amount of current passing through the coil. This is illustrated by Fig. 1-3.

Q1-1. What are the ammeter ranges in the meter shown in Fig. 1-2?

Q1-2. What is the maximum voltage measurement?

Q1-3. How many ohmmeter ranges does it have?

Q1-4. What is the maximum resistance reading?

Your Answers Should Be:

A1-1. 0–10 mA, 0–250 mA, and 0–500 mA.

A1-2. 500 volts.

A1-3. Three.

A1-4. Infinity.

Meter Torque

When a small current passes through the coil, a weak magnetic field is produced. This causes a small turning force (torque) to exist between the coil field and the permanent field. Thus, the coil and pointer rotate a small amount. A larger current through the coil produces a stronger magnetic field around the coil, a greater torque, and more rotation of the coil and pointer.

Meter Coil

The meter coil is formed of fine wire wrapped on a rectangular aluminum frame. The coil frame is mounted so that it can rotate freely in the gap between the core and the poles. In some meters, a screw on the front panel (just above the pointer axis) permits accurate adjustment of the pointer position. The pointer should read zero (ammeter and voltmeter scales) when no current is flowing through the coil.

Meter Sensitivity

Meter sensitivity is expressed in two ways—**current** sensitivity and **ohms-per-volt** sensitivity.

Current Sensitivity—This is determined by the amount of current required by the meter movement to cause a full-scale deflection of the pointer. Current sensitivity depends on the number of turns in the meter coil. It also depends on the strength of the permanent-magnet field. Current sensitivity is expressed as the number of milliamperes (mA) or microamperes (μA) required for full-scale deflection of the meter needle. Typical meter movements have current sensitivities of 1 mA and 50 μA. This is illustrated in Fig. 1-4.

Ohms-per-Volt Sensitivity—Ohms-per-volt sensitivity is determined by the total resistance that must be in series in the meter circuit to obtain full-scale deflection when 1 volt is ap-

Fig. 1-4. Current sensitivity.

plied. The resistance can be calculated by using Ohm's law:

$$R\ (\text{ohms/volt}) = \frac{E\ (1\ \text{volt})}{I\ (\text{current sensitivity})}$$

For a 50-μA (0.00005 ampere) meter, the resistance is 1/0.00005, or 20,000 ohms, resulting in a sensitivity of 20,000 ohms-per-volt.

Fig. 1-5. Ohms-per-volt sensitivity.

Q1-5. What causes the meter pointer to move?

Q1-6. If the permanent magnetic field between the pole pieces decreases in strength, the meter will give readings that are (more than, less than, the same as) those given before.

Q1-7. Which meter will have a greater number of turns in its coil, a 40-μA or a 2-mA movement?

Q1-8. If the coil rotates, how does current get from the meter circuitry to the coil?

Q1-9. What is the ohms-per-volt sensitivity of a meter with a current sensitivity of 2 mA?

Q1-10. If the resistance in a dc voltmeter circuit is 10,000 ohms for full-scale deflection at 1 volt, what value of resistance must be substituted to measure 10 volts full scale?

Your Answers Should Be:

A1-5. **Torque,** resulting from the reaction between the magnetic fields of the coil and the pole pieces, causes the meter pointer to move.

A1-6. A decrease of the magnetic field strength of the pole pieces will cause the meter to give readings **less than** those given before.

A1-7. The **40-μA meter movement** will have a greater number of turns in its coil.

A1-8. **Calibrated springs,** one at each end of the coil, form the current path to the coil.

A1-9. A meter movement of 2 mA has a sensitivity of **500 ohms-per-volt.**

A1-10. **100,000 ohms.**

Meter Circuitry

An analog meter is a current-reading device. To provide accurate readings, the electrical values of its circuit components must be fairly precise. The circuit design must provide for all of the types of measurements to be made by the meter.

If a meter measured only one characteristic (voltage, current, or resistance), and if it had only a single range (10 V, 1 mA, or 1000 ohms, for example), the design would be relatively simple. Without considering ranges, there are four basic types of circuits found in multimeters.

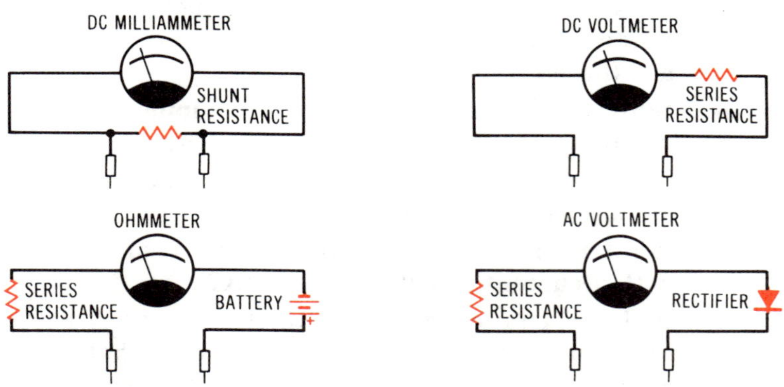

Fig. 1-6. Multimeter circuits.

Meter-Circuit Components

Fig. 1-6 shows that multimeter circuitry requires series and/or shunt resistances, a battery for measuring ohms, and a diode for limiting current direction when measuring ac voltage. To obtain different ranges of volts, amperes, or ohms, resistances of selected values must be used in the circuit. The circuits are connected to front-panel controls that readily provide a means of selecting the desired function and range.

Rotary Wafer Switches—These switches are often used to provide range selection. As seen in the schematic diagram of Fig. 1-7, a metallic wafer can be rotated to one of several positions. The blade of the wafer engages taps, or contacts, that are connected to appropriate parts of the circuit.

A multimeter may have only a single rotary switch with enough wafers to select both the function (ohmmeter, milliammeter, dc voltmeter, or ac voltmeter) and the appropriate range for each function. In a multimeter having two rotary switches, one switch usually selects the meter function and the other selects the range.

Q1-11. In the schematic of Fig. 1-7, to which resistance is the positive test lead connected?

Q1-12. What multimeter function does the circuit provide?

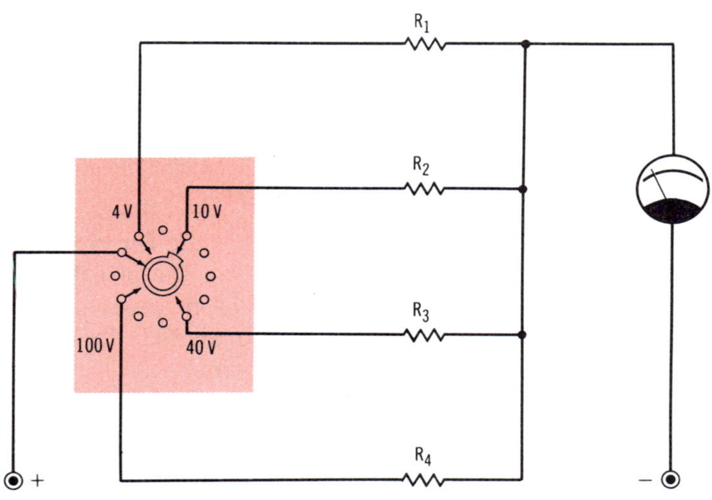

Fig. 1-7. Circuit with a wafer switch.

Your Answers Should Be:

A1-11. The test lead is connected to $\mathbf{R_2}$.

A1-12. The switch and circuitry indicate that the circuit can be used to measure **dc volts.**

Ganged Rotary Switches—The schematic in Fig. 1-8 shows another method of representing rotary switches. It indicates how a pair of switches (S_1 and S_2) might be used to make all the necessary connections in a multimeter. The dashed line shows that the S_2 switch sections are **ganged.** That is, the wafers are mechanically connected to the same shaft and stacked in decks along the shaft.

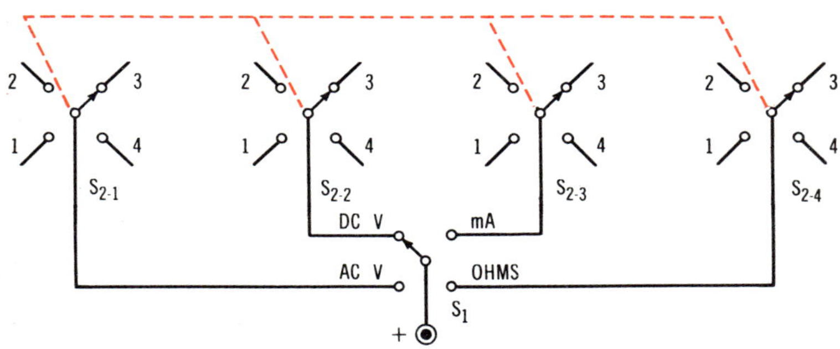

Fig. 1-8. Ganged rotary switches.

METER FRONT PANEL

The front panel of a multimeter has a scale for reading values, and provision for setting or adjusting the position of the pointer. A means of selecting the type of measurement to be taken, and the desired ranges, is also provided. Jacks for inserting test leads are mounted on the panel.

One means of selecting measurement and range scales was shown in Fig. 1-8—rotary wafer switches. In some multimeters, a combination of **pin jacks** and a rotary switch is used. The switch selects the desired range. The red multimeter test lead is inserted in a pin jack marked with the quantity to be measured, and the black test lead is placed in the pin jack marked COMMON or – (minus).

Another method, often used, employs a number of pin jacks located on the front panel. Test leads are inserted in the desired positions—one in the jack marked COMMON and the other in the jack marked with the desired measurement. A circuit diagram suitable for dc voltage and current measurements, using this arrangement, is shown in Fig. 1-9.

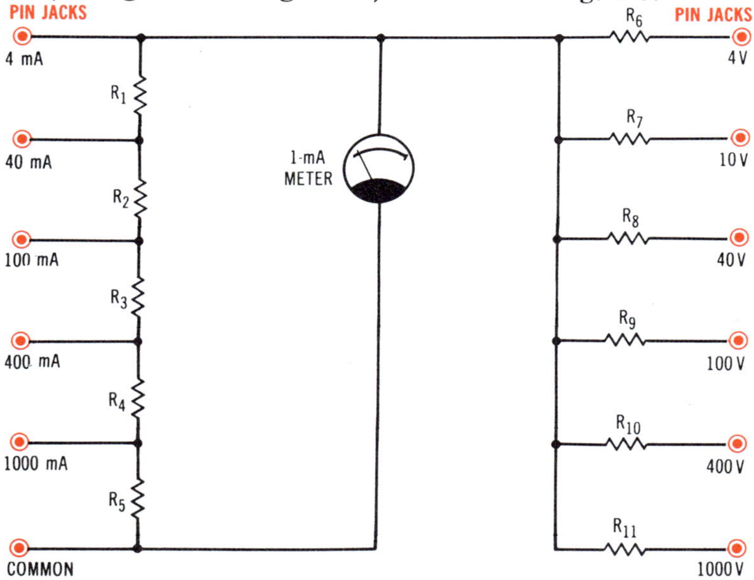

Fig. 1-9. Multimeter with pin jacks.

Whatever the arrangement might be, always check the settings before taking a measurement. If the switch is positioned in the wrong function, or in too low a range, the meter could be damaged.

Q1-13. If the test leads were in the COMMON and 400-mA jacks in Fig. 1-9, what resistances would be shunting the meter?

Q1-14. With the test leads in the COMMON and 1000-V jacks, current will flow through the meter and which resistances?

Q1-15. With the voltage values shown, which resistance would have the larger value, R_8 or R_9?

Your Answers Should Be:

A1-13. $\mathbf{R_4}$ and $\mathbf{R_5}$ would be shunted across the meter. R_1, R_2, and R_3 would be in series with the meter.

A1-14. Current will flow through $\mathbf{R_1}$, $\mathbf{R_2}$, $\mathbf{R_3}$, $\mathbf{R_4}$, $\mathbf{R_5}$, and $\mathbf{R_{11}}$.

A1-15. $\mathbf{R_9}$ would be larger than R_8.

CURRENT AND VOLTAGE SCALES

Scales on an analog multimeter are usually calibrated to measure the quantities marked on the selection switches or jacks. A single scale can be used for more than one function and range. If a separate scale were used for each type of measurement and for each range, the meter face would be cluttered and difficult to read. The types of measurements you have learned about thus far can be made using a meter face with either two or three scales.

The sketch shown in Fig. 1-10 illustrates a two-scale meter —one scale for ohms and the other for ac–dc voltage and dc current. The lower scale is calibrated so that each mark has two values. The value to use depends on the selected range. A meter with three scales would probably have separate scales for ohms, dc volts/milliamperes, and ac volts. However, each of the separate scales for the different functions may have more than one set of numbers for the divisions on the scales. Thus, it is possible to read values on more than one range for each function.

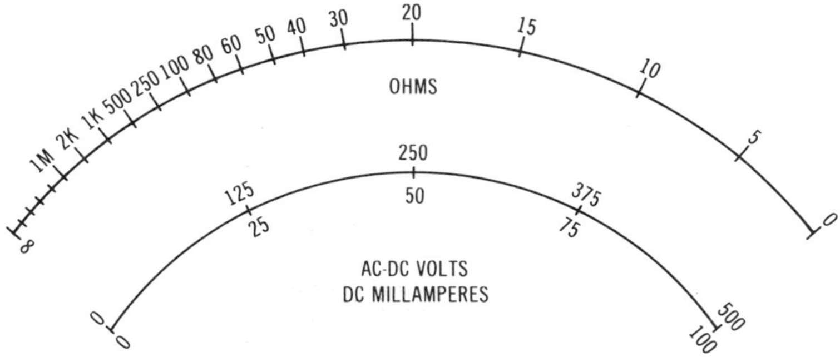

Fig. 1-10. Example of a meter face having two scales.

Multimeter scales for reading voltage or current are usually linear. This means that the divisions on the scale are spaced equal distances apart. On a scale that measures from 0 to 100, for example, the halfway mark would be 50. Midway between 0 and 50 is 25. If major divisions are marked off in smaller units, the spaces between subdivisions are also equal.

Reading Linear Scales

A linear scale is not difficult to read if care is taken. To keep the scale uncluttered, only the major divisions are numbered. If the pointer rests between numbers or marks, the correct quantity can easily be estimated by determining the units in which the subdivisions are calibrated.

The major divisions are 0, 25, 50, 75, and 100 on the scale in Fig. 1-11. The magnified portion shows subdivisions of five units each. There are additional marks halfway between these subdivisions.

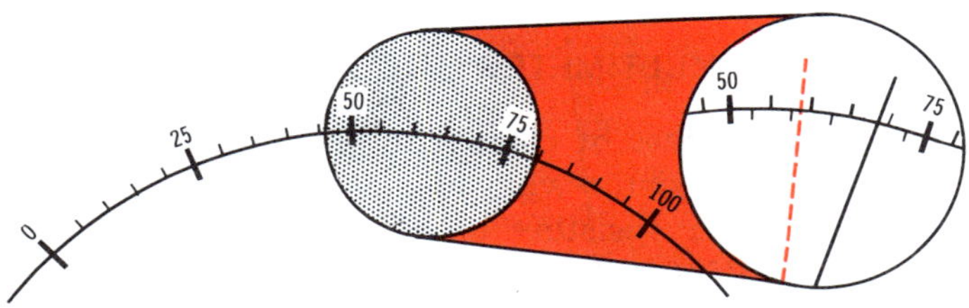

Fig. 1-11. Sketch for Problem Q1-16.

Q1-16. The solid pointer in the diagram of Fig. 1-11 is between 65 and 70 and slightly beyond the midmark. The meter reads 68. What is the reading indicated by the dotted pointer?

Q1-17. The solid pointer reads 78 in the diagram of Fig. 1-12. What does the dotted pointer read?

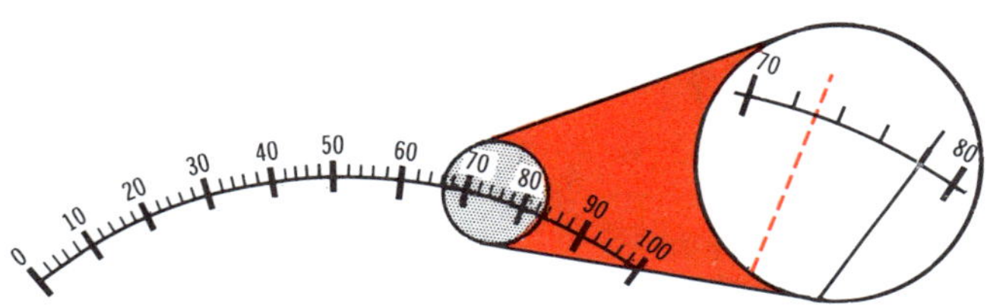

Fig. 1-12. Sketch for Problem Q1-17.

Your Answers Should Be:

A1-16. 58.

A1-17. 73.

Linear-Scale Markings

There is no standard system for marking linear scales. The left end of the scale is usually zero. On rare occasions, however, zero may be on the right end. This only means the pointer will move from right to left. The other end may be 10, 15, 25, 40, 50, 60, 100, or some other number. To use the scale, determine the quantity contained between numbered markings and the values of the indicated subdivisions. You should have no trouble if you make this determination with care. Estimating readings between markings will be no more difficult than the readings you made on the preceding page.

MULTIMETER ACCURACY

The electrical values of components are never precise. You have learned that resistors vary as much as ±20% of the stated ohmic value. Better resistors have tolerances of 10 and 5%. The tolerance rating of the resistors used in the multimeter affects the accuracy of the meter readings. In Fig. 1-13, a 100,000-ohm resistor was selected to give a full-scale meter reading of 100 volts. If the resistor had a tolerance of 20%, the full-scale reading could be off about 20 volts in either direction. A 10% resistor could cause readings between the extremes of about 90 volts and 110 volts. A 5% resistor could result in an error of 5 volts above or below 100 volts. None of these readings would be close enough for most purposes.

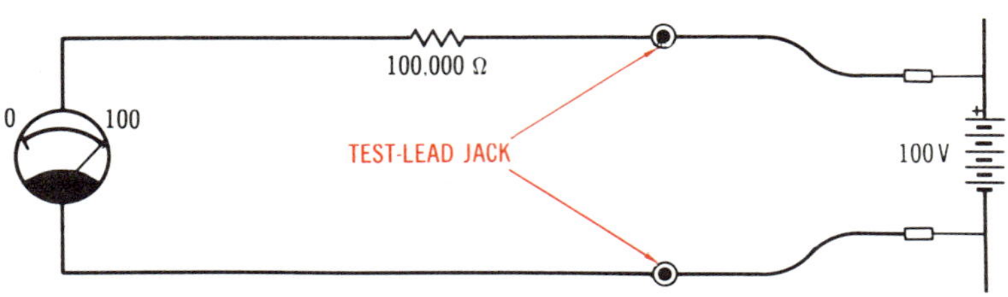

Fig. 1-13. Circuit used to get full-scale meter deflection.

Most multimeters employ ±1% resistors. In the example given, a 100-volt reading would not be off more than 1 volt, a 1% error. This is close enough for most measurements.

The meter movement itself may give some error. For example, its readings might be off 1 volt throughout the 100-volt scale. One volt off at 100 volts is a 1% error, but 1 volt at 10 volts is a 10% error. For this reason, voltage and current readings should be taken on the upper half of the scale.

LINEAR-SCALE RANGES

It is sometimes necessary to take measurements as high as 100 volts or even 1000 volts; at other times, a reading of only a few volts will be called for. A 1000-volt scale would provide any of these readings, but the accuracy of a 6-volt measurement would be very poor. Also, small differences on a 1000-volt scale would be difficult to read. To overcome these limitations, multimeters have several ranges.

Multiple-Range Scale Reading

The scale shown in Fig. 1-14 is marked in divisions from 0 through 10. When the range selector is set for 10 V, the exact measurement is read directly from the actual scale markings.

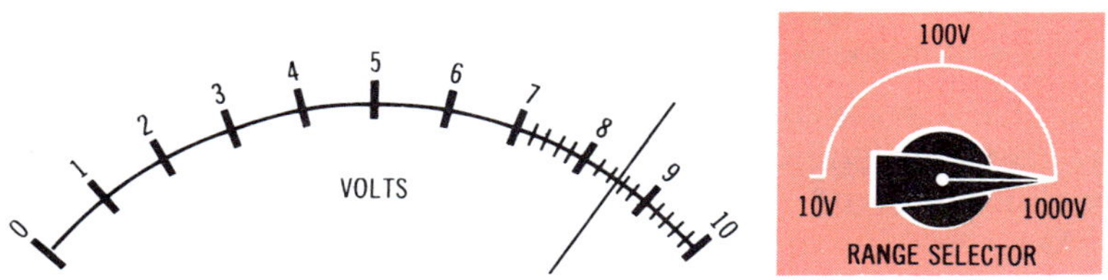

Fig. 1-14. Sketch for Problem Q1-18.

If the selector is set on 100 V, the scale readings are multiplied by 10 to obtain the measured value. If 10 volts were the measured value, the pointer would show a full-scale reading if the meter were on the 10-V range. On the 100-V range, the pointer would come to rest over the 1 mark—10 times 1 equals 10 volts.

Q1-18. What voltage is being measured with the pointer and range settings shown in Fig. 1-14?

Q1-19. A multimeter has ranges of 10 V, 50 V, 100 V, and 500 V. Which range should be used to read 45 V?

Your Answers Should Be:

A1-18. 850 V.

A1-19. The 50 V range.

Dual-Marked Scales

Another example is a single scale having two values for each of its markings (Fig. 1-15). This type is used in multimeters whose ranges are not multiples of ten times a single full-scale quantity. Study the scale and range settings shown in Fig. 1-15. The principles used in the preceding example still apply. With the range setting shown, the meter reads 32.5 volts.

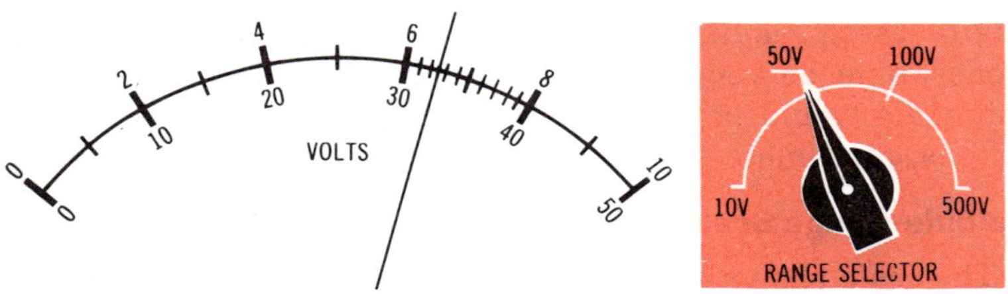

Fig. 1-15. Dual-marked scales.

Multimeasurement Scales

A linear scale can be used to measure more than one electrical characteristic. The preceding examples used voltage readings. These could have been either ac or dc volts as far as the scale or range settings were concerned. The same scale can also be used to read milliamperes. Appropriate switches on a multimeter might look like the examples shown in Fig. 1-16.

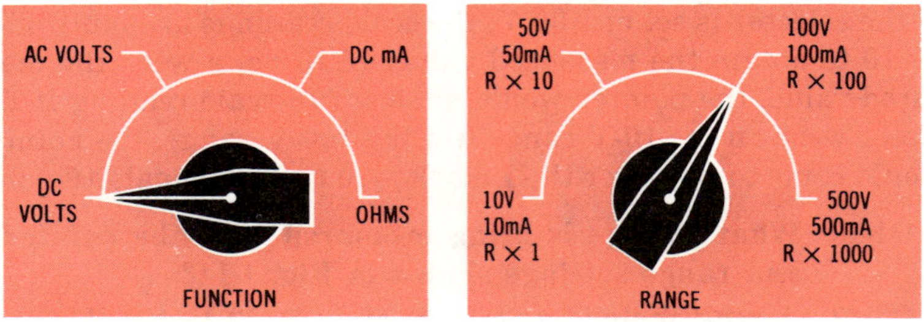

Fig. 1-16. Multimeter switches.

Measurement Precautions

When making a measurement, form the habit of taking the reading first on the highest-value range. Such a precaution can prevent damage to the meter. For example, a 500-volt measurement taken at a 10-volt range setting will cause excessive current to pass through the meter coil. This will result in either burning out the coil or bending the pointer against the retarding pin. Therefore, unless you are absolutely certain of what range to use, set the multimeter for the highest-value range when taking the first reading.

OHMMETER SCALES

The scale and range selection for the resistance-reading portion of the multimeter are a little different from those already discussed. The scale reads from 0 to infinity (∞) instead of from 0 to some number. Unlike the volt and milliampere scales, the resistance scale is not linear. Range selection is indicated by multipliers (R $\times$ 1, R $\times$ 10, R $\times$ 100, etc.) instead of a quantity indicating full-scale deflection. These differences will become apparent as you examine the basic ohmmeter circuit shown in Fig. 1-17.

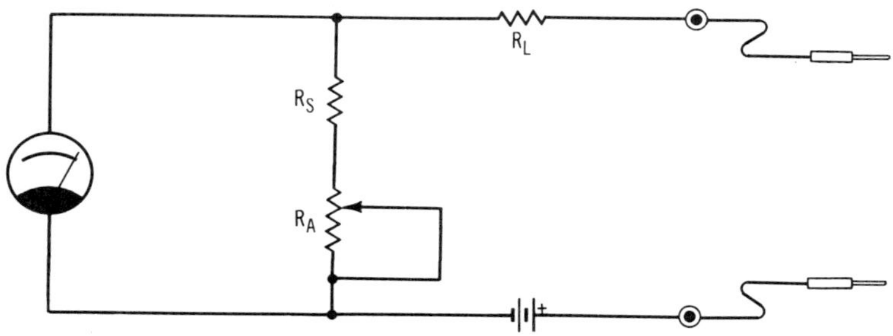

Fig. 1-17. A basic ohmmeter circuit.

Q1-20. On which range should a meter be set when making the first measurement of a circuit quantity?

Q1-21. Your meter has range settings of 10 V, 50 V, 100 V, and 500 V. You wish to measure the voltage across a load and suspect a voltage of 90 V. How should you set your meter to read this voltage?

Your Answers Should Be:

A1-20. When taking the first reading, set the meter on the **highest range** for the meter function being used.

A1-21. The meter should first be set on the **500 V range.** This setting will ensure against meter damage.

Ohmmeter Circuits

An ohmmeter circuit must supply its own source of current. Usually a self-contained battery is used for this purpose. The voltage of the battery is determined by the sensitivity of the meter, the arrangement of the series and shunt resistors in the circuit, and the size of the external resistance to be measured. Depending on the design of the ohmmeter, the battery might be from 1.5 to 45 volts.

Resistance R_L in the circuit in Fig. 1-17 is a current-limiting resistance in series with the meter. Its value is determined by the amount of current required to cause full-scale deflection. Resistors R_A and R_S form a shunt across the meter. Therefore, only a fraction of the total current in the circuit flows through the meter. The current through the meter is determined by the ratio of the meter resistance to the shunt resistance. Resistance R_A, controlled by the zero-ohms-adjust knob on the panel, establishes the value of total shunt resistance that will cause the meter to register accurate readings.

Determining Ohmmeter-Scale Markings

In the diagram shown in Fig. 1-18, resistance values in the parallel network—R_M (meter resistance), R_S (shunt resistance), and R_A (zero-ohms adjust)—are such that a full-scale pointer deflection will occur when 1 mA enters the network. If the battery voltage is 1.5 V and the circuit resistance is 1500 ohms, 1 mA ($I = E/R$) will flow when the test leads are shorted (touched together). The meter will show full-scale deflection of the pointer, or zero ohms. When the test leads are parted, no current flows, and the pointer returns to its normal position. The ohmic reading becomes infinity (∞). This is the reason for zero being at the right-hand end on most ohmmeter scales.

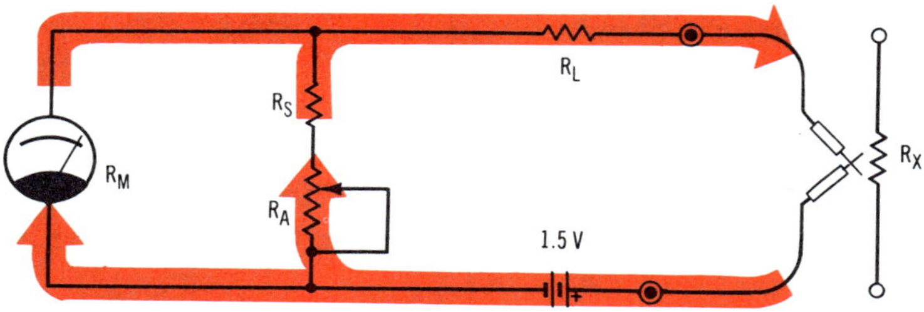

Fig. 1-18. Circuit for determining ohmmeter-scale markings.

Using Ohm's law, plot the value of current and the resulting scale positions when resistances (R_X) of 500, 1500, and 4500 ohms are measured. If R_X is 500 Ω and the resistance of the ohmmeter circuit is 1500 Ω, the total resistance is 2000 Ω.

$$I = \frac{1.5\ V}{2000\ \Omega} = 0.00075 \text{ Amp, or } 0.75 \text{ mA}$$

As far as the meter is concerned, it will receive the same ratio (or fraction) of any current flowing into the parallel network. Since 1 mA is required for full-scale deflection, 0.75 mA will move the pointer to three-fourths of full scale. Calculating current for the other values of resistance, you should be able to plot a chart that looks like the following listing.

R_X	R_L	R_T	I_T	Scale Deflection
0	1500	1500	1 mA	Full
500	1500	2000	0.75 mA	¾
1500	1500	3000	0.50 mA	½
4500	1500	6000	0.25 mA	¼
Inf.	1500	Inf.	0 mA	Zero

Q1-22. What factors determine the voltage of an ohmmeter battery?

Q1-23. What factors determine the fraction of total current that flows through the meter coil in an ohmmeter circuit?

Q1-24. The zero reading is usually on the _____ end of an ohmmeter scale.

Your Answers Should Be:

A1-22. The **sensitivity of the meter, arrangement of the shunt and series resistors,** and **size of the external resistance** to be measured.

A1-23. The **ratio of meter resistance to shunt resistance.**

A1-24. The zero reading is usually on the **right** end of an ohmmeter scale.

Ohmmeter-Scale Design

Starting at the right, compare deflection and ohmic readings at the quarter points on the ohmmeter scale shown in Fig. 1-19. The first quarter covers 500 ohms, the second quarter equals 1000 ohms (1500–500), the third is 3000 ohms, and the fourth is infinity. Such a scale cannot be calibrated in linear divisions.

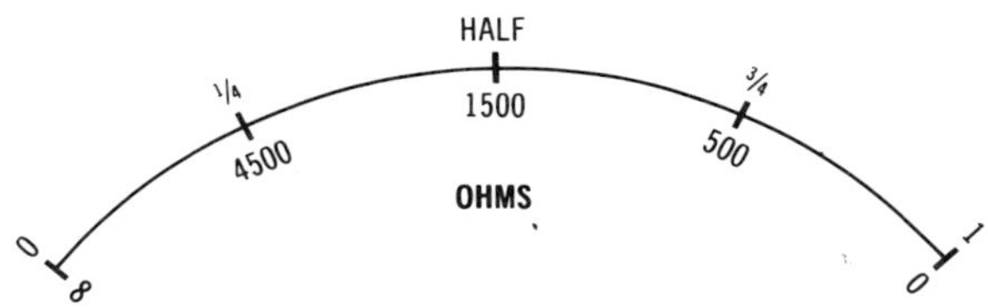

Fig. 1-19. Basic ohmmeter-scale design.

Reading Accuracy

Because of the nonlinearity of ohmmeter scales, readings should be taken with the pointer in the most readable area of the scale. A rule used by many technicians is to read values in the area of the scale bounded by 1/10 and 10 times the value of the midscale reading. However, if only one range is available, such a rule is not practical. For example, if a meter reads 10 ohms at midscale, all resistance measurements would have to fall between 1 and 100 ohms. This is not possible. Therefore, several ohmic ranges are provided in a multimeter.

Resistance Ranges

Typical ohmmeter ranges are R × 1, R × 100, and R × 10K. Some multimeters have multipliers as high as R × 10 million. Using the rule mentioned above, the R × 1 range provides low resistance readings (0 to 100 or 200 ohms). The R × 100 range will give useful readings between 100 ohms and 10K, and the R × 10K range will be satisfactory for readings from 10K to 1 megohm. Higher readings may be estimated with fair accuracy. If an R × 1M range is available, resistances up to about 100 megohms can be measured.

USING THE OHMMETER

The basic procedures for measuring resistance are the same for all multimeters. First, set the meter to read **OHMS.** Plug the test leads into the proper jacks. Hold the tips of the test probes together, thus placing a short (zero ohms) across the internal circuit. Turn the **zero-ohms adjust** (sometimes labeled **zero adjust**) control until the meter pointer rests at zero on the ohms scale. Each time the meter is set for reading ohms or it is switched to a new range, short the test probes together and zero the pointer with the **ohms adjust** control.

> **CAUTION**
>
> Never use an ohmmeter to take a resistance reading across an energized circuit. The internal circuit is designed to carry only the current developed by its own battery. Its voltage is usually between 1.5 and 9 volts. Voltage from an external source will usually be larger than this value and will damage either the meter coil, the pointer, or some internal part.

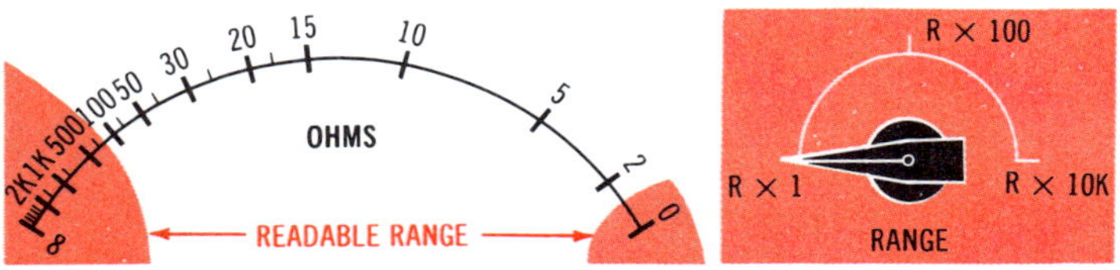

Fig. 1-20. Sketch for Problem Q1-25.

Q1-25. Using the diagram shown in Fig. 1-20, which range setting should be used to measure an estimated 1500 ohms?

Your Answer Should Be:
A1-25. The range selector should be set on **R × 100.**

MULTIMETER CIRCUITS

Understanding how function and range circuits work is important in learning how to use a multimeter properly. Studying these circuits will also help develop your skill in analyzing electronic circuits. For these reasons, multimeter circuits will be studied a portion at a time. As each section is discussed, it will be added to the preceding portions until a typical multimeter circuit is developed. Each portion will be diagrammed first using a rotary range switch and, then, as it might appear employing pin jacks.

Milliammeters

The schematic diagram given in Fig. 1-21 shows a typical dc milliammeter circuit with rotary switch connections. Resistor values are not shown. They will vary according to the multimeter.

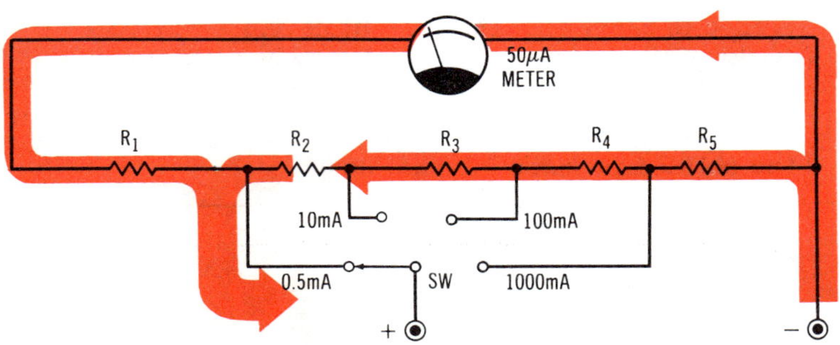

Fig. 1-21. A dc milliammeter circuit.

Notice that the arrangement is a parallel circuit in which the resistance can be changed in both branches by rotating a switch. Since the meter has a 50-μA movement (full-scale deflection), the ratio of resistances must be such that 50 μA will be the maximum current flowing in the meter branch at each switch setting. Redrawing the circuit for each switch position may help make this clear.

How much of the total current (0.5 mA) must flow through the shunt? A current of 500 μA (0.5 mA) minus 50 μA (maximum meter current) equals 450 μA. This means that the ratio

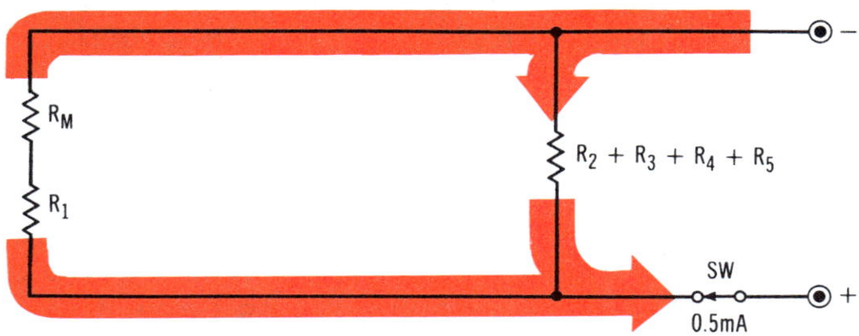

Fig. 1-22. Circuit for 0.5-mA switch position.

of shunt current to meter current must be maintained at 9 to 1. Therefore, the shunt resistance must be 1/9 of the resistance in the meter branch. With some meters, resistor R_1 is included as a part of the basic meter movement to increase its resistance. This is sometimes necessary so that the shunt resistance for the high-current ranges will not be unreasonably small.

Q1-26. What is the ratio of shunt current to meter current in the circuit of Fig. 1-23?

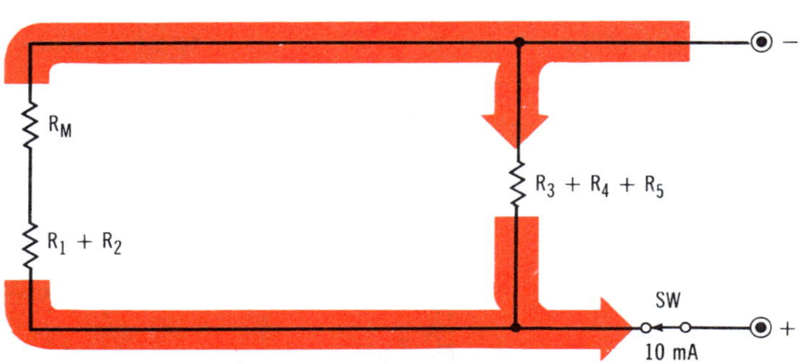

Fig. 1-23. Circuit for 10-mA switch position.

Your Answer Should Be:

A1-26. $\frac{\text{I (shunt)}}{\text{I (meter)}} = \frac{9{,}950\ \mu\text{A}}{50\ \mu\text{A}} = \frac{199}{1}$

Shunt Ohmmeters

In the illustration given in Fig. 1-24, R_M represents the meter resistance. Resistance R_A is the **ohms adjust** control and is used to set the pointer to zero reading (full-scale deflection). In the R × 1 range, R_M, R_A, R_6, R_7, and R_8 form one branch of a parallel network; R_9 forms the other.

The shunt permits external resistances of low values to be read with reasonable accuracy. Without it, the measured resistance would be in series with the resistance in the meter branch, and the same current would flow through all resistances. The total resistance would, thus, have to be large to limit current flow to the maximum level of 50 μA. As a consequence, slight changes of current due to small changes in measured resistances would cause only tiny changes in pointer movement. Markings would be very close together.

An external resistance equal to the internal circuit resistance would bring the pointer to midscale. With a 50-μA meter and a 1.5-volt battery, Ohm's law shows that the internal resistance must be 30,000 ohms for full-scale deflection, or zero reading. This 30,000 ohms of external resistance would halve the current flow and would provide a midscale reading. On the R × 1 range, measurements between 0 and 30,000 ohms would be distributed on one half of the scale and readings would be difficult to estimate with accuracy.

Placing a shunt in the circuit provides a low-resistance cur-

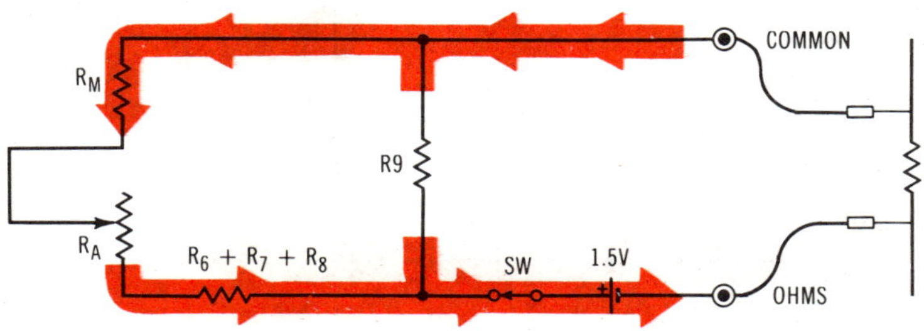

Fig. 1-24. Circuit for the R × 1 meter range.

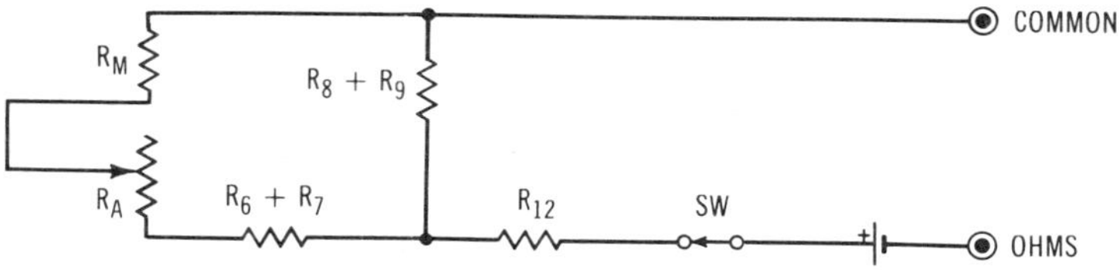

Fig. 1-25. Circuit for the R × 10 meter range.

rent path around the meter branch. Most multimeters using this design have a midscale reading of 10 to 30 ohms, permitting a greater accuracy in reading the scale markings.

A typical value for R_9 is 11 or 12 ohms. In round figures, the total resistance of the parallel network would then be about 10 ohms. With the terminals shorted, the 1.5-volt battery will cause a total current (I_T) of 0.15 ampere (150,000 μA) to flow. Since maximum current for the meter is 50 μA, one part in 3000 (50/150,000) of the total current will flow through the meter branch. The remainder (2999 parts) flows through resistor R_9, the shunt path. Resistance R_A is used to adjust the meter branch resistance to produce a 2999/1 ratio.

When the range-selector switch is moved from R × 1 to R × 10, the resistances in the parallel network are redistributed. Compare the schematic given in Fig. 1-25 with the R × 1 circuit in Fig. 1-24. The resistance is increased in the shunt branch and decreased in the meter branch. Resistor R_{12} is added as a current-limiting resistance.

Resistor R_{12} in the circuit of Fig. 1-25 will allow less current to flow than is in the circuit of Fig. 1-24. Maximum I_T in the R × 1 circuit would be near 150,000 μA if the parallel network had a resistance of 10 ohms. Suppose the resistance of the parallel network of the R × 10 circuit (Fig. 1-25) is approximately 100 ohms and R_{12} is 10 ohms. The 1.5-volt battery is then in a circuit having a total resistance (R_T) of 110 ohms. The total current, with test leads shorted, would approach 15,000 μA. Since I_M must be 50 μA, 50/15,000 or 1/300 of the total current flows through the meter.

Q1-27. In which circuit (Fig. 1-24 or Fig. 1-25) will the meter pointer be farther from zero when a given resistance is measured?

Your Answer Should Be:

A1-27. The circuit of **Fig. 1-24.**

Redrawing the two circuits with meter scales added will help you understand why the above answer is correct. Resistor R_8

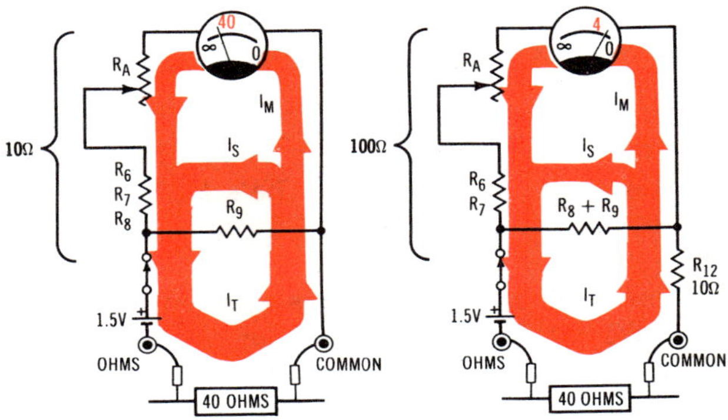

(A) R × 1. (B) R × 10.

Fig. 1-26. Circuits for meter ranges of R × 1 and R × 10.

is in the meter branch for Fig. 1-26A and in the shunt branch for Fig. 1-26B. The ratio of meter-branch to shunt-branch resistance for Fig. 1-26B is smaller than for Fig. 1-26A. This means that if I_T were equal in both parallel networks, I_M for Fig. 1-26B would be larger than the I_M for Fig. 1-26A. Therefore, a larger part of the total current would flow through the meter in Fig. 1-26B than through the meter in Fig. 1-26A.

R × 1 Circuit:

Total circuit R = 50 ohms
Total circuit I = 30,000 μA (I = E/R)
I_M/I_S = 1/3000 (meter-branch to shunt-branch ratio)
I_M = 10 μA (1/3000 of 30,000 μA)

R × 10 Circuit:

Total circuit R = 150 ohms
Total circuit I = 10,000 μA (I = E/R)
I_M/I_S = 1/300 (meter-branch to shunt-branch ratio)
I_M = 33.3 μA (1/300 of 10,000 μA)

There will be a greater deflection (more I_M) in the circuit of Fig. 1-26B than in the circuit of Fig. 1-26A when the same value of resistance is measured. The proportions established for the series-parallel resistances in the R × 10 circuit are such that scale readings can be multiplied by 10.

On the R × 100 and R × 1000 meter ranges, the series-parallel resistive networks of the circuits are changed in the same manner, and the scale readings are multiplied by factors of 100 and 1000, respectively. In the R × 1000 circuit shown in Fig. 1-27, you will note that there is an additional battery so as to permit measurement of large resistances.

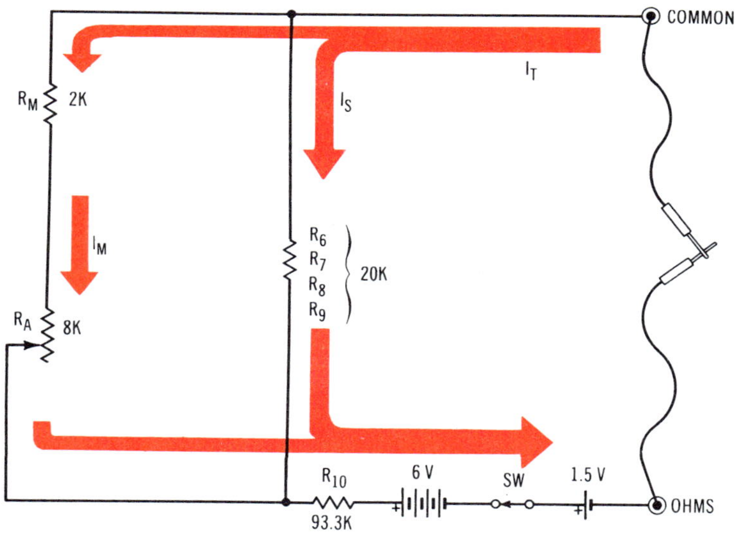

Fig. 1-27. Circuit for the R × 1000 range.

Q1-28. What is the value of R_T in the circuit of Fig. 1-27?

Q1-29. What is the purpose of the additional 6-volt battery in the meter circuit (Fig. 1-27)?

Q1-30. What is the value of I_T in the circuit when the test leads are shorted?

Q1-31. What is the ratio of meter-branch current to shunt-branch current?

Q1-32. How much of the total current will pass through the meter?

Your Answers Should Be:

A1-28. R_T is equal to **99.97K.**

$$R_T = \frac{(2K + 8K) \times 20K}{(2K + 8K) + 20K} + 93.3K = 99.97K$$

A1-29. The additional 6-volt battery **permits large resistances to be measured accurately.**

A1-30. I_T is equal to **approximately 75 μA.**

$$I_T = \frac{E}{R_T} = \frac{7.5\ V}{99.97K} = 75\ \mu A$$

A1-31. $I_M/I_S = 2/1$

A1-32. $I_M = 50\ \mu A$ **(⅔ of 75 μA)**

Dc Voltmeter Circuits

Fig. 1-28 shows a typical dc voltmeter circuit added to the circuit of the dc milliammeter and ohmmeter. As you can see, the dc voltmeter circuit is a network of four resistors in series

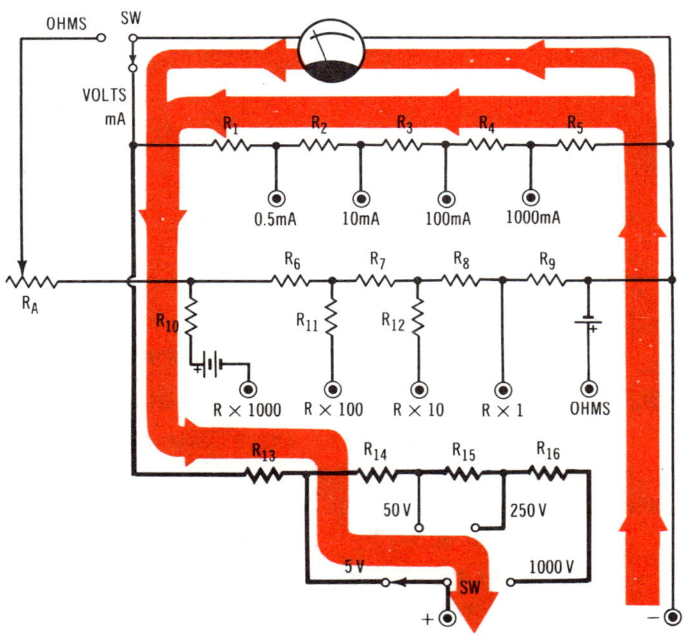

Fig. 1-28. A dc voltmeter circuit.

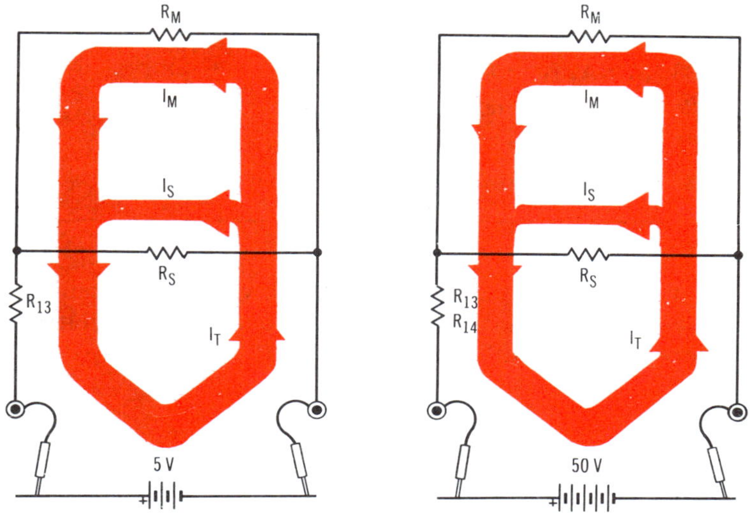

Fig. 1-29. Circuits for 5-volt and 50-volt ranges.

with the meter-shunt resistor combination. Simplified schematics for the 5- and 50-volt positions are shown in the diagrams of Fig. 1-29.

In the 5-volt position, current flows through R_{13} and the meter-shunt network. The 50-μA meter has a coil resistance of 2000 ohms. Since R_S is very close to the same value, it can be assumed that the total resistance of the parallel network is 1000 ohms. What must R_{13} be for a full-scale reading of 5 volts? If R_M and R_S are equal, I_M and I_S are also equal. Therefore, for I_M to be 50 μA, I_T must be 100 μA. R_T equals 5 V/100 μA, or 50,000 ohms. Subtracting 1000 ohms, R_{13} is 49,000 ohms. To find the value of R_{14}, apply the same reasoning.

Q1-33. If 3.7 volts is measured using the 5-volt circuit, how much current will flow through the meter?

Q1-34. If R_{14} decreases in value, will the meter for the 50-volt range read high or low?

Q1-35. What is the value of R_{15} (250-volt circuit in Fig. 1-28)?

Q1-36. What is the value of R_{16} (1000-volt circuit in Fig. 1-28)?

Your Answers Should Be:

A1-33. **37 μA.** $I_M = \frac{1}{2} I_T$ **(74 μA)**

A1-34. The meter will read **high.**

A1-35. **2 megohms**

A1-36. **7.5 megohms**

Ac Voltmeters

The diagram in Fig. 1-30 shows an ac voltmeter circuit. The four multiplier resistances are connected as in the dc voltmeter, but there are other differences.

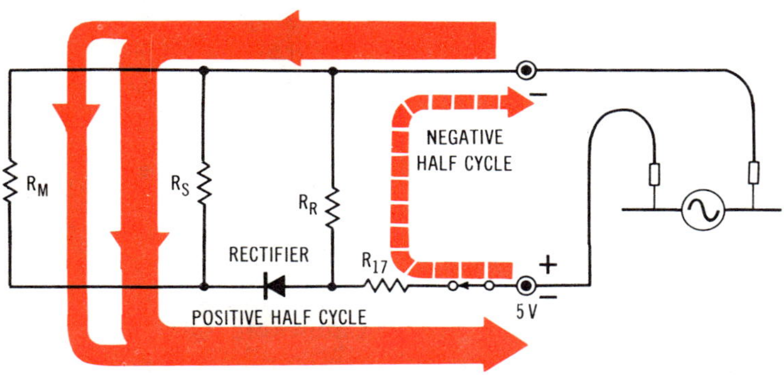

Fig. 1-30. An ac voltmeter circuit.

When the polarity of the measured voltage is positive at the plus terminal, current will flow as indicated. It will pass through the rectifier. When the voltage swings in the negative direction, current flow will reverse. The rectifier offers a high resistance to current in this direction. Therefore, only a very small amount of current can flow through the meter during this part of the cycle. Consequently, the meter receives a pulsating dc, and the pointer is deflected. If an alternating current passed through the meter in both directions, the pointer would remain at zero.

The rectifier resistance is low in comparison to resistance R_R when the current flows counterclockwise, but it is high with respect to R_R when the current flows in the opposite direction.

Resistor R_R becomes a path for current during the negative half cycle, thereby preventing a possibly destructive negative voltage from building up across the rectifier.

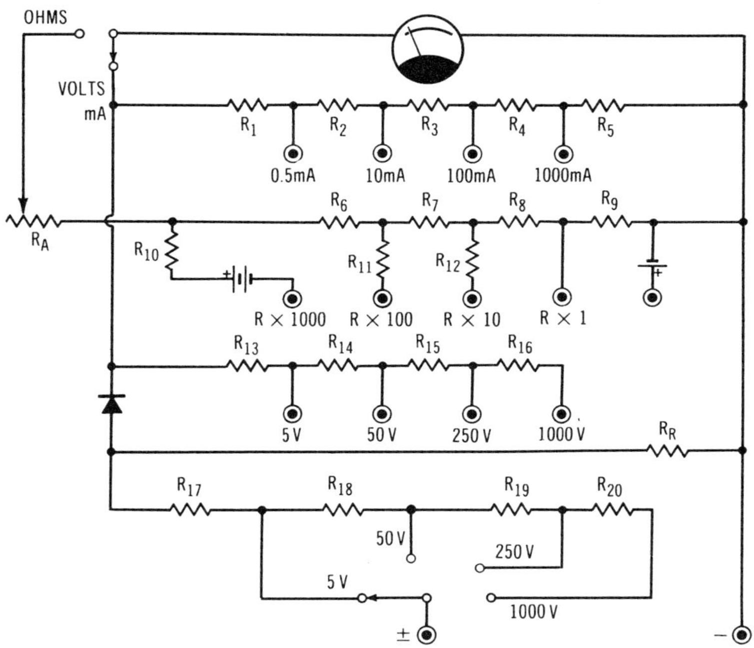

Fig. 1-31. A multimeter circuit.

USING A MULTIMETER

You have learned a great deal about why an analog multimeter operates as it does. This knowledge should aid you in using one wisely. However, it might be helpful to list the more important **do's** and **don'ts.**

General Precautions

Before using any multimeter, carefully study and apply the information contained in its instruction book. If the book does not contain a schematic diagram of the circuitry, request one from the dealer or manufacturer. Fig. 1-31 gives an example of a typical circuit. Study the diagram and learn how the circuits are connected.

Keep the front panel clean. Dirt or moisture around the jacks may act as a shunt for current. Although it may look rugged, handle the instrument with care.

Always take readings with as much precision as possible. Develop the accuracy habit early. Then, when you need to, you will be able to take precise measurements.

Be continually alert to the possibility that the equipment you are working on may be "hot" to ground and, therefore, the source of a dangerous shock. The chassis of some radios and tv sets are examples.

Handle the front-panel controls carefully. Do not try to rotate switches beyond their stops.

Keep your hands away from the metal tips of the test probes. Failure to do so may cause you to receive an electrical shock when measuring current or voltage. Also, the resistance of your body across the probes will make ohmmeter readings inaccurate.

Voltmeter and Milliammeter Functions

Have a great respect for an energized circuit. Stand on dry insulating material, and if you must measure voltage of great amounts, (1) turn the equipment off, (2) discharge any capacitors near the test point, (3) clip the meter leads onto the test points, connecting the common lead first, and then the "hot" lead, (4) turn the equipment on, (5) take the meter reading, and (6) turn the equipment off before removing the meter test leads.

Never place the milliammeter circuit across a voltage source. Even a small amount of voltage may force an excessive amount of current through the meter coil. To measure current, always connect the milliammeter in **series** with the circuit. Always turn the equipment off before removing the meter from the circuit.

Always connect the voltmeter in parallel with the circuit, voltage source, or circuit component.

Observe polarity. Place the negative test probe (usually black) on the negative side of the element and the positive probe (usually red) on the positive side.

Ohmmeter Function

Do not measure resistance in an energized circuit. Turn off the appropriate switches, remove the power plug, disconnect the battery terminals, or take any other measure that will remove voltage from the circuit. A very small voltage added to that of the ohmmeter battery may damage the meter coil plus, possibly, other components.

Discharge any capacitors in and around the circuit before making a resistance reading. Remember, capacitors store voltage. The ohmmeter may serve as a discharge path for a capacitor when the meter leads make connection in the circuit. The metal shaft of a screwdriver may be used to discharge low-voltage capacitors. Rub the blade against all leads or terminals of the capacitor while the shaft rests on the chassis. Hold the screwdriver by its handle. For high-voltage capacitors (300 V or above), use a **grounding tool.** Fasten the clip of the grounding tool (Fig. 1-32) to the bare metal of the chassis. Touch the terminals or leads of the capacitor with the tip of the rod.

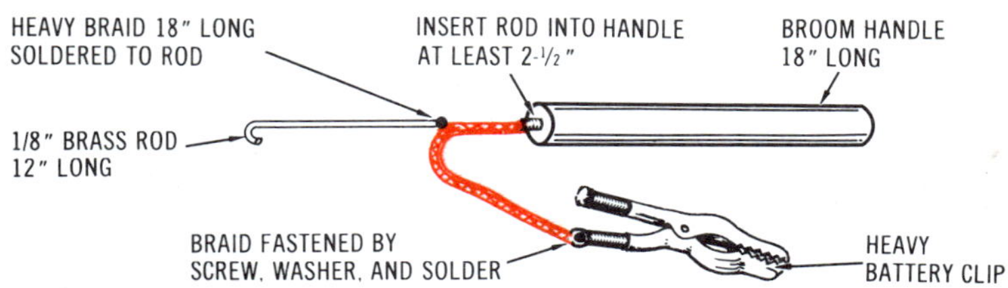

Fig. 1-32. A grounding tool.

When taking resistance readings in a circuit, try to determine if the element being measured is shunted by another component. If it is, the reading may be affected. If such a condition exists, you have two choices; either remove one of the component leads from its terminal before measuring, or use the point-to-point resistance values contained in the instruction book or technical manual for the equipment under test. The resistance values may be shown in chart form. One type of chart that lists the resistance values between tube pins and chassis ground is shown by the following example.

Tube	Pin 1	Pin 2	Pin 3	Pin 4	Pin 5	Pin 6	Pin 7	Pin 8	Pin 9
V_1	Inf	615K	0	0.1	0	6.2K	165K	Inf	0
V_2	10.3K	68	1 meg	100	Inf	16K	1.5K		
V_3	22K	100	920	0	100	100	0	1	
etc.									

* Resistance values are given in ohms.

Do not measure the resistance of circuit elements that are still hot. Readings taken on parts above room temperature may be inaccurate.

Another type of chart may be just for a single circuit, but it gives both voltage and resistance readings from the tube pins to chassis ground. The following table of voltages and resistances is an example of such a chart.

Element	Pin No.	Voltage	Resistance
Plate	1	320 V ac	280 Ω
Filament	3	3.15 V ac	0
Filament	4	3.15 V ac	0
Plate	6	320 V ac	280 Ω
Cathode	7	+350 V	125K

When using these charts, be sure the front-panel controls of the equipment are set as indicated by the manual containing the chart. Also, be sure to use a meter with the same sensitivity as the type that was used by the manufacturer in developing the chart.

If your measurements are the same or reasonably close, those components included in the reading may be good. You cannot be absolutely sure, however. For example, a normally high-value resistor may be open. If this resistor is in parallel with low-resistive components, such as a coil, the ohmmeter reading may agree with the chart value. A leaky capacitor across a resistor may also produce a normal reading. However, if you suspect this condition or get an abnormal reading, disconnect one of the leads of the suspected component and make another measurement.

The condition of capacitors can be approximated with an ohmmeter. When testing capacitors other than electrolytics, use the highest resistance range of the meter. This range will supply more voltage than the others. If the capacitor is good, the meter pointer will deflect slightly and, then, return to infinity as the capacitor charges from the ohmmeter battery. If there is no deflection, the capacitor may be open or have too small a value for the size of the ohmmeter battery. Full-scale deflection (zero resistance) with no return indicates a shorted capacitor. Leakage is indicated by a steady deflection to some part of the scale.

WHAT YOU HAVE LEARNED

1. A multimeter is an instrument used to measure ohms, ac and dc volts, and dc milliamperes.
2. A combination of switches and jacks on the front panel of a multimeter permits the instrument to be used to measure these electrical characteristics using just a single meter.
3. Meters vary in sensitivity. Sensitivity can be stated in two ways—in current and in ohms-per-volt. Current sensitivity, rated in milliamperes or microamperes, indicates the amount of current flow through the meter coil necessary to cause a full-scale deflection of the meter pointer. Current sensitivity of most multimeters ranges from 2 mA to 50 μA. The smaller current rating means greater sensitivity. Ohms-per-volt sensitivity is determined by the amount of meter-circuit resistance that will result in full-scale deflection when 1 volt is applied to the meter leads. A 2-mA current sensitivity would be rated at 500 ohms-per-volt ($R = E/I$). A 50-μA meter movement would have a sensitivity of 20,000 ohms-per-volt. The latter meter is preferred, since it adds less loading effect to a circuit being measured.
4. Most multimeters have a variety of ranges for each of the four meter functions—ohms, dc volts, ac volts, and dc milliamperes. Ranges are obtained by selecting internal circuit arrangements through the use of switches or jacks.
5. The circuit arrangements are parallel or series-parallel resistive networks. Each range circuit sets up a distribution of resistances in the meter and shunt branches

of the network. The resistance ratios are such that no more than the maximum meter current will flow through the meter branch of the network. The excess current is diverted through the shunt branch.

6. Full-scale deflection of the pointer (meter needle) occurs when maximum rated current flows through the meter coil.
7. When the instrument is set up as an ohmmeter, full-scale deflection occurs when the test leads are shorted. This indicates ZERO ohms.
8. The voltmeter and ammeter scales on the meter face are usually linear. Units of scale markings are equal distances apart. The same scale can be used for measuring both volts and milliamperes. On some multimeters, a separate scale for each is available.
9. The ohmmeter scale is nonlinear—markings are not equal distances apart. The scale reads from zero (usually at the right end) to infinity.
10. For best accuracy, all multimeter readings should be made in the range position where the pointer will be in the upper-half region of the scale. There is some error present in even the best meter movement. This difference will be less of a measurement error at full scale than near zero. Also, the markings of the ohmmeter scale are more crowded toward the zero end of the scale (full-scale deflection).

2

Vacuum-Tube Voltmeters

what you will learn

In this chapter, you will learn the functions and operations of a vacuum-tube voltmeter (vtvm). A vtvm is an electronic analog voltmeter. You will learn, by comparison, the advantages of a vtvm versus those of a multimeter. After completing this chapter, you will know which type of meter to use for any circuit.

CAPABILITIES

A vtvm, because of its design, has a few advantages not found in multimeters. The primary advantage is that it can be used for measuring voltages without excessively loading a

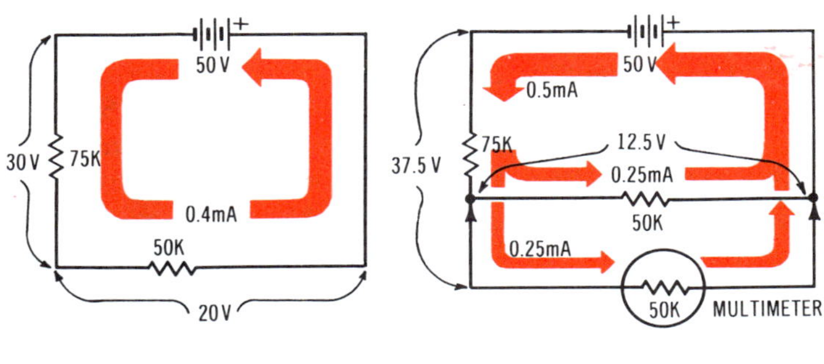

Fig. 2-1. A multimeter loads a circuit.

circuit. For example, a 1000-ohm-per-volt multimeter that is set on its 50-volt range will place 50,000 ohms across the circuit being measured. Since this 50,000-ohm resistance provides another path for the circuit current to follow, the multimeter is said to be **loading** the circuit.

According to Ohm's law, you should be able to measure 20 volts across the 50K resistor shown in the left-hand circuit in Fig. 2-1. A 1000-ohm/volt meter set to the 50-volt range will place 50,000 ohms across the resistor, thus loading the circuit. The voltmeter will, therefore, read only 12.5 V (0.25 mA × 50K). This is a large error. A 20,000-ohm/volt multimeter, because of its higher input resistance, causes less circuit loading. A typical vtvm has a 10,000,000-ohm input resistance, regardless of range. Study the following chart, and compare the loading effects of the two meters.

Range	Input Resistance		Loading Effect
	VTVM	Multi*	
5 V	10 meg	0.1 meg	Vtvm is 100 times less.
10 V	10 meg	0.2 meg	Vtvm is 50 times less.
50 V	10 meg	1 meg	Vtvm is 10 times less.
100 V	10 meg	2 meg	Vtvm is 5 times less.
500 V	10 meg	10 meg	Multimeter is equal to VTVM.
1000 V	10 meg	20 meg	Multimeter is 2 times less.

* Multimeter with 20,000-ohm/volt sensitivity.

As you can see, a vtvm has less loading effect on circuits at the lower voltages than does a good multimeter. Some circuits are so sensitive to loading effects that a reading can be obtained only with a vtvm. Another advantage of the vtvm is its wider frequency coverage. The ac voltage-reading error in a multimeter increases at the rate of 0.5 to 1% per each 1000 cycles of frequency. At 5000 hertz (cycles per second), for example, a multimeter would read 2.5 to 5% low. At 50,000 hertz, readings would be 25 to 50% low. A vtvm, however, will provide ac measurements of reasonable accuracy up to tens of megahertz.

VTVM'S

There are many different designs for constructing a vtvm. Most of them use some type of vacuum-tube bridge circuit to regulate the amount of current that flows through the meter

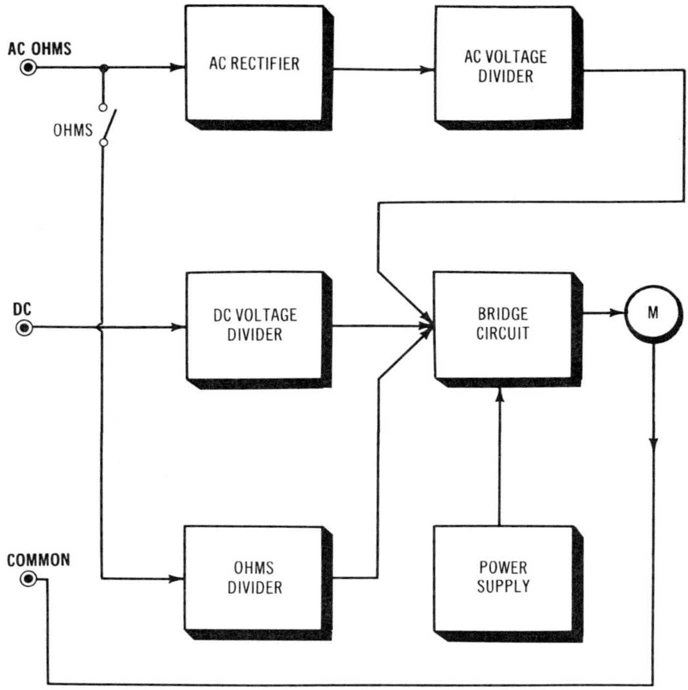

Fig. 2-2. Block diagram of a typical vtvm.

coil. This method not only provides an accurate means of measurement, but it also permits the use of a less sensitive meter. A typical vtvm might use a 200-μA meter movement, whereas, a 20,000-ohm/volt multimeter would require a more expensive 50-μA movement.

The block diagram of a vtvm, given in Fig. 2-2, shows three input jacks for test leads. By following the arrows, you can see that when measuring alternating current, the input current flows through a rectifier, a voltage divider, a bridge circuit, and the meter before it returns through the COMMON pin jack. Current, representing direct current and ohmic readings, flows through the respective divider networks in the circuit to the bridge circuit and the meter. From there it returns to the COMMON jack.

Q2-1. Why does a vtvm provide more accurate readings than a multimeter when measuring low voltages?

> **Your Answer Should Be:**
>
> **A2-1.** A vtvm has a **greater impedance** across its input jacks at lower voltage ranges than a multimeter. Therefore, it **draws less current** from the circuit.

VACUUM-TUBE VOLTMETER CIRCUITS

In the diagram of Fig. 2-3, the inputs are shown in a block rather than as individual jacks. This vtvm would have four jacks on the panel, each labeled with the function titles shown. The function switch (S_1) connects the jacks to the proper

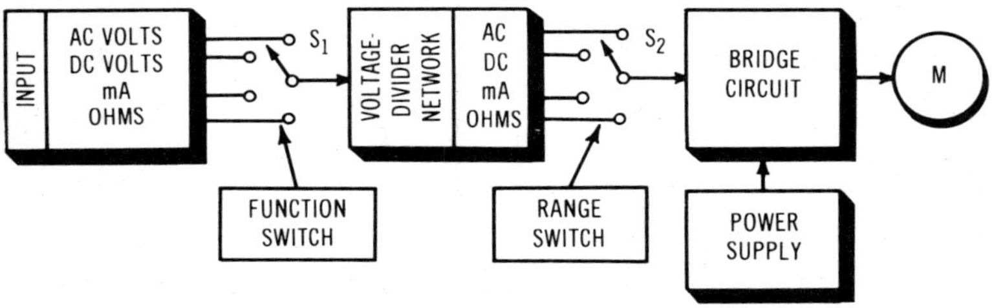

Fig. 2-3. Pictorial block diagram of a vtvm.

voltage-dividing networks. The range switch (S_2) selects the proper network, usually resistive, to supply the bridge circuit with the correct voltage. The bridge circuit provides the amount of current that should flow through the meter for the correct pointer deflection. The power supply provides both filament and B+ voltages to the vacuum tubes.

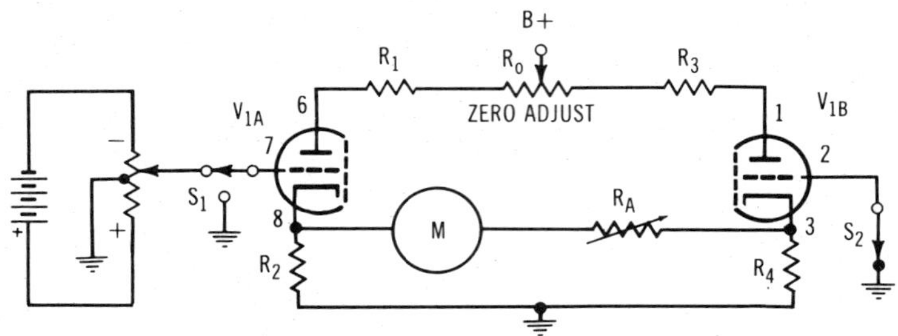

Fig. 2-4. A bridge circuit.

Bridge Circuits

The bridge circuit will be described first since it represents the most significant difference between a vtvm and a multimeter. The schematic given in Fig. 2-4 shows a typical circuit. As the schematic shows, V_{1A} and V_{1B} are separate halves of a twin triode. (The A and B designations represent separate groups of elements in the same tube envelope.) The arrangement of the circuit is such that the plate current of V_{1A} can be made equal to the plate current in V_{1B}. The balance in plate currents is brought about by the adjustment of R_0. The control for R_0 is on the vtvm front panel and is labeled **ZERO ADJUST.** Before taking a measurement, this control is adjusted until the pointer is resting on zero on the desired scale.

A pointer reading of zero means no current is flowing through the meter. This condition exists when the two cathode voltages are the same with respect to ground. Since cathode resistors R_2 and R_4 are equal in value, the voltages are equal as long as the two plate currents remain the same.

A 200-μA meter is sufficiently sensitive to operate in such a circuit. Potentiometer R_A is a range-calibrating resistor. When the range switch is moved to a new position, a different R_A is switched into the meter circuit.

If the switch (S_1) at the grid (pin 7) of V_{1A} is thrown from ground to the battery and resistance circuit, a negative voltage will be applied to the grid. Plate current will decrease by an amount determined by the change in grid voltage. A decrease in plate current will lower the voltage across resistor R_2. Since the voltage across resistor R_4 has not changed, a difference of potential exists across the meter. This potential difference results in a current flow through the meter.

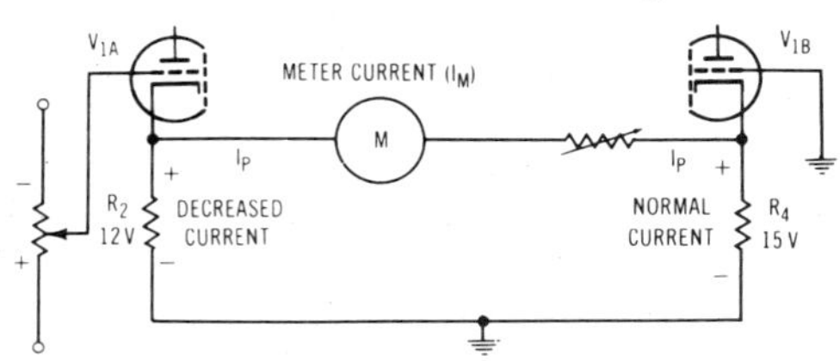

Fig. 2-5. Circuit for Problem Q2-2.

Q2-2. In which direction will meter current flow in the circuit shown in Fig. 2-5?

Your Answer Should Be:

A2-2. The meter current will flow from **left to right.**

Assume that a normal plate-current flow through V_{1B} produces a cathode voltage of 15 volts. Also, assume that the decrease in plate current through V_{1A} drops its cathode voltage to 12 volts. Current flows through the meter from the less positive to the more positive side.

If the voltage applied to the grid of V_{1A} represents a full-scale reading at the range setting in use, what should be the value of R_A? A 200-μA meter has a full-scale deflection when a current of 200 μA flows through its coil. If the voltage across the meter is 3 volts, the resistance of R_A plus the meter resistance must be 15,000 ohms.

$$R = \frac{E}{I} = \frac{3\text{ V}}{0.0002\text{ A}} = 15{,}000\text{ ohms} = 15\text{K}$$

Fig. 2-6 shows the circuit redrawn in the form of a bridge.

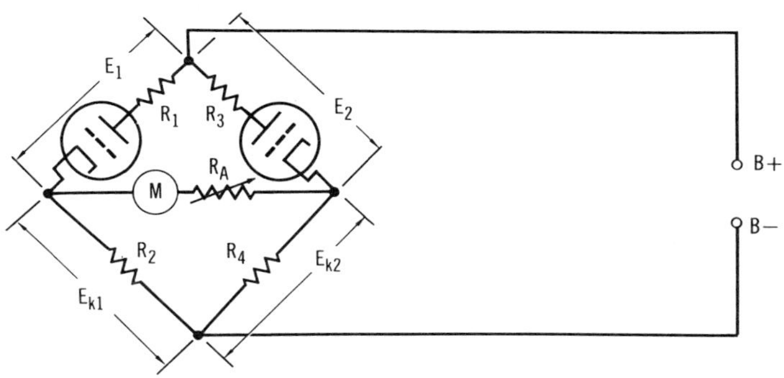

Fig. 2-6. Circuit in Fig. 2-5 redrawn as a bridge circuit.

With no voltage applied to the grid of either tube, the current is the same on both sides of the bridge. Cathode voltages E_{K1} and E_{K2} are equal. If a negative voltage appears on the left grid, current through the left leg of the bridge decreases. The voltage drop across resistor R_1 decreases, but the voltage across the tube increases enough that voltage E_1 increases. However, the decrease in current through R_2 lowers the value of E_{K1}, and the sum of E_{K1} and E_1 remains the same. The bridge is unbalanced because E_{K1} is less than E_{K2}. Current through the meter is from left to right. If the negative voltage is removed from the grid on the left side of the bridge, the bridge will return to its normal balanced condition. Voltages on both sides will be equal, and no current will flow through the meter.

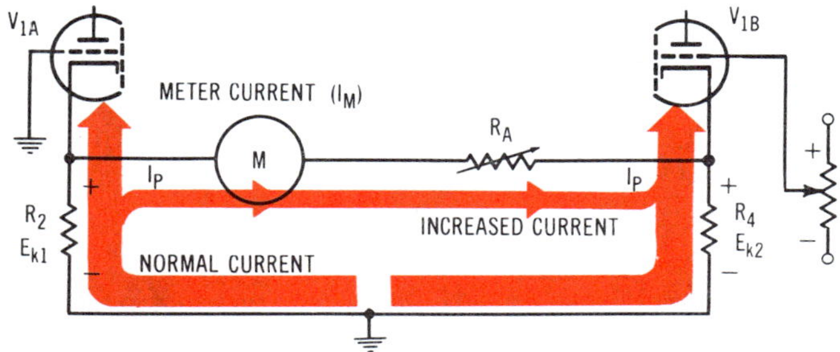

Fig. 2-7. Positive voltage is applied to one grid.

Suppose a positive voltage is applied to the grid of V_{1B} in Fig. 2-7. Plate current in V_{1B} will increase, raising the voltage across R_4. E_{K2} will now be a larger positive voltage than E_{K1}. Current will, therefore, flow left to right through the meter.

Importance of Chassis Potential

You have seen that a negative voltage applied to one grid, or a positive voltage to the other, results in the same direction of current flow through the meter. This fact is used to great advantage in a vtvm.

The common, or negative, test-lead jack is grounded to the chassis of a vtvm. In equipment employing electronic circuits, the chassis is usually at zero (ground) potential for safety purposes. If you have to measure a negative voltage in such equipment, the test leads must **not** be interchanged in order to cause the meter pointer to move up scale. Placing the common, or negative, lead on the negative side of the voltage would connect this voltage directly to the vtvm chassis and case. This could be very dangerous.

Q2-3. What voltage across a meter leg would produce full-scale deflection of a 200-μA meter? Assume the total leg resistance is 20K.

Q2-4. What conditions must exist to get a zero deflection on the meter in the bridge circuit shown in Fig. 2-6?

Your Answers Should Be:

A2-3. $E = I \times R = 0.0002\ A \times 20{,}000\ \Omega = 4$ **volts.**

A2-4. E_{K1} **must equal** E_{K2}. Therefore, R_2 **must equal** R_4, and the **plate currents must be equal.**

Measuring Negative and Positive Voltages

A vtvm eliminates the danger of having the wrong chassis potential by having dc− and dc+ settings on the function switch (Fig. 2-8). The type of bridge circuit that was just

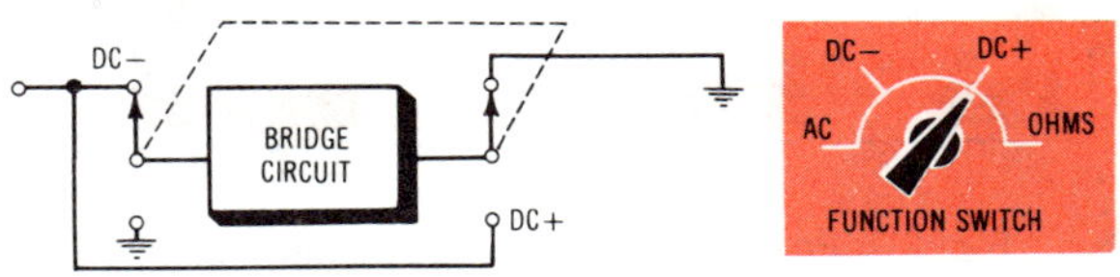

Fig. 2-8. A dc reversing switch.

discussed allows the input voltage to be switched to the grid of the proper tube in the circuit. Thus, the same direction of current flow through the meter can be maintained regardless of the polarity of the input voltage.

Dc Voltage Divider

In a vtvm, the input circuits for the various functions are voltage dividers. A typical dc voltage divider is shown in the diagram of Fig. 2-9. One end of the string of resistors is attached to the dc pin jack, and the other end is connected to

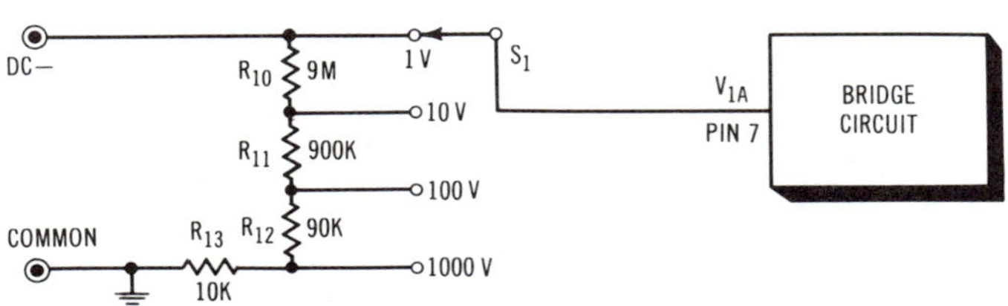

Fig. 2-9. A dc voltage-divider circuit.

the COMMON jack. A schematic representation of a four-position switch is shown at the right of the voltage divider. Each position represents a tap in the resistor string. Each tap corresponds to a range setting.

The sum of the resistances in the voltage divider is 10 megohms. This large resistance will be in parallel with the measured circuit at any range setting. Loading of the circuit under test is therefore very small (Fig. 2-10).

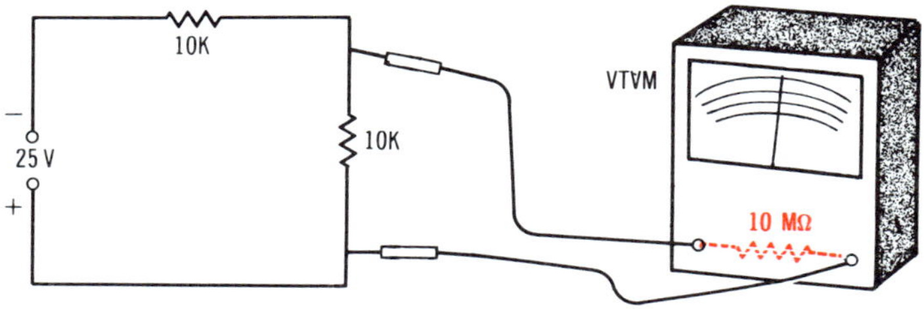

Fig. 2-10. A vtvm produces a small loading effect.

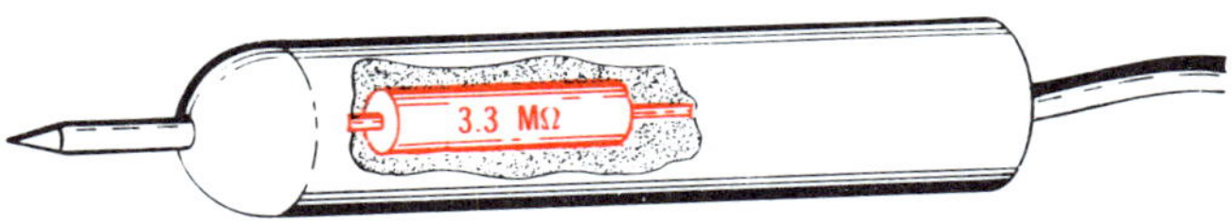

Fig. 2-11. A dc test probe.

Q2-5. In the diagram of Fig. 2-12, how much voltage appears on the grid of the tube in the bridge circuit?

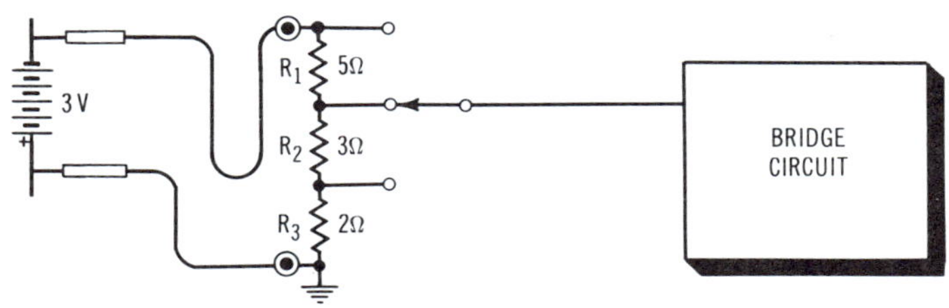

Fig. 2-12. Circuit for Problem Q2-5.

Q2-6. Refer to the illustration of Fig. 2-9. How much voltage is applied to pin 7 of V_{1A} when 1 volt is measured with S_1 positioned as shown?

Q2-7. How much voltage is applied to pin 7 of V_{1A} when 10 volts is measured with S_1 switched to the 10-volt tap of the voltage-divider circuit?

Your Answers Should Be:

A2-5. There is **1.5 volts** applied to the grid.

$$I = \frac{E}{R_T} = \frac{3\ V}{10\ \Omega} = 0.3\ \text{ampere}$$

$$I \times (R_2 + R_3) = 0.3\ A \times 5\ \Omega = 1.5\ V$$

A2-6. **1 volt.**

A2-7. **1 volt.** A voltage of 10 volts is applied across a total resistance (R_T) of 10M. At the 10-volt tap, the resistance to ground is 1M, or 1/10 R_T. Therefore, 1 volt is present at the 10-volt tap.

Dc Voltmeter Probes

Two different probes for taking measurements are available with a vtvm. The standard probe has a resistor in series with the metal point and the test lead (Fig. 2-11).

High-Voltage Probes

A high-voltage probe has an internal series resistor of 25 megohms or more. This probe is used to extend the range of the vtvm above 1000 volts dc (Fig. 2-13).

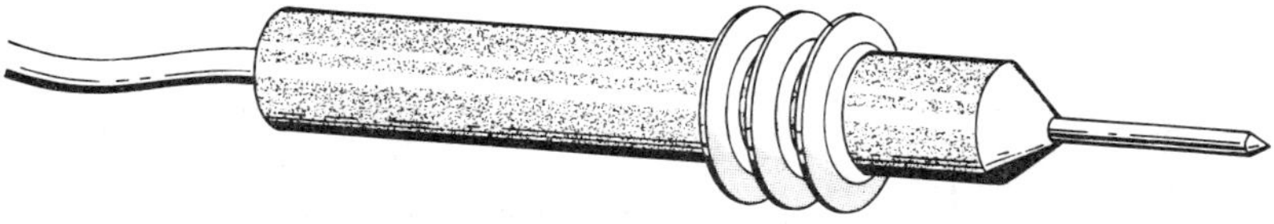

Fig. 2-13. A high-voltage dc probe.

Voltage-Divider Principles

Answers A2-6 and A2-7, above, illustrate an important point. With the test probes measuring 1 volt, there is 1 volt at the 1-volt tap. When measuring 10 volts, there is 1 volt at the 10-volt tap. And, at 100 and 1000 volts, there is 1 volt at the 100-volt and 1000-volt taps, respectively. The voltage applied to the bridge circuit must be the same at all ranges to cause the same pointer deflection. Pointer positions are multiplied by the factor indicated by the range setting.

Ac Voltage Dividers

A simplified schematic for an ac voltage divider is shown in Fig. 2-14. It has the same voltage-divider network used in the dc voltmeter circuit. The main difference between this circuit and the dc circuit is a pair of twin diodes for rectifying ac into a pulsating direct current. The diodes permit the measurement of voltages having frequencies from 50 Hz to at least 50 kHz (kilohertz). Diodes CR_1 and CR_2 and capacitors C_5 and C_6 form a network that converts the ac input voltage into a dc voltage across the voltage divider. Capacitor C_5 also serves to prevent the dc that may be present in the input voltage from reaching the bridge circuit. In this arrangement, the vtvm actually measures the peak value of the ac voltage. The **rms value** of a sine wave is 0.707 times the peak value. The scales of most vtvm's are marked to read the rms value.

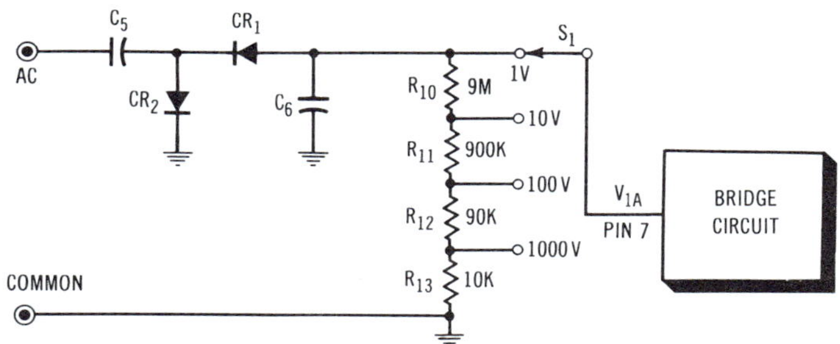

Fig. 2-14. An ac voltage-divider circuit.

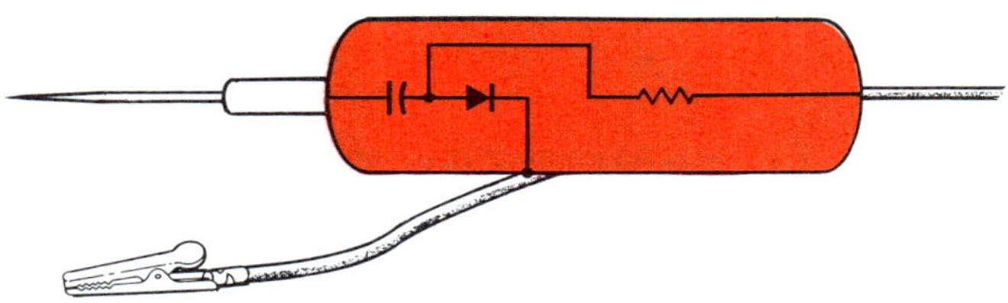

Fig. 2-15. An rf probe.

Q2-8. In addition to being part of the network that converts ac to dc, what is the purpose of capacitor C_5 in the ac voltage-divider circuit of Fig. 2-14?

Q2-9. What is the purpose of the two diodes in the ac voltmeter circuit?

Your Answers Should Be:

A2-8. Capacitor C_5 is also used as a **blocking capacitor** to prevent the dc component of any measurement from entering the ac measuring circuit.

A2-9. The two diodes **rectify the incoming ac** into pulsating dc.

Rf Probes

A special probe made for measuring frequencies up to 100 megahertz (MHz) can be used in the dc jack; such a probe is illustrated in Fig. 2-15.

Ohmmeter Voltage Divider

Some ohmmeter circuits employ a battery as a voltage source. In others, a metallic rectifier with a resistor-capacitor filter is used. Whatever the source of voltage, the current normally flows through a voltage-divider network similar to the one shown in Fig. 2-16.

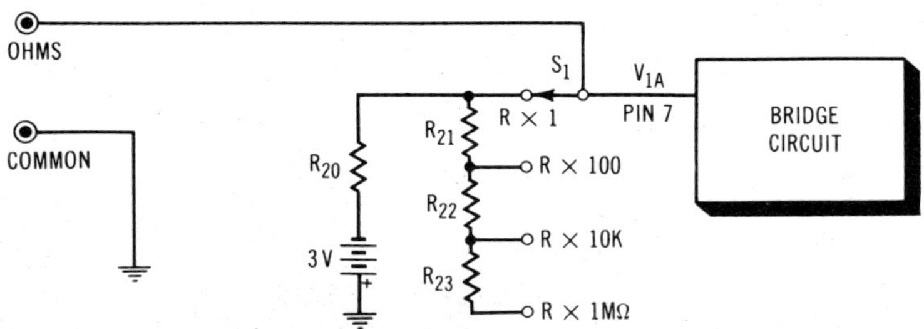

Fig. 2-16. An ohmmeter voltage divider.

With S_1 in the R × 1 position, the voltage applied to the bridge circuit depends on the ratio of the unknown resistance to R_{20}. In the R × 100 position, the voltage depends on the ratio of the unknown resistance to $R_{20} + R_{21}$. The unknown resistances measured on the R × 100 range are 100 times as large as those measured on the R × 1 range. You can see that $R_{20} + R_{21}$ must be 100 times as large as R_{20}. Similar reasoning applies to the R × 10K and R × 1 MEG ranges.

The ohmmeter scale of a vtvm, using this circuit, measures 0 to ∞ from left to right—the reverse direction to that of a typical multimeter scale. When the probes are shorted, full-scale deflection will occur. When they are open, the pointer will remain stationary and register ∞. As in the multimeter, the ohmmeter scale is nonlinear.

USING THE VTVM

The same precautions given for the use of multimeters also apply to vacuum-tube voltmeters. Generally, a vtvm is used in the same manner as a multimeter when taking voltage and resistance readings. The illustration in Fig. 2-17 shows how the front panel of a typical vtvm might appear.

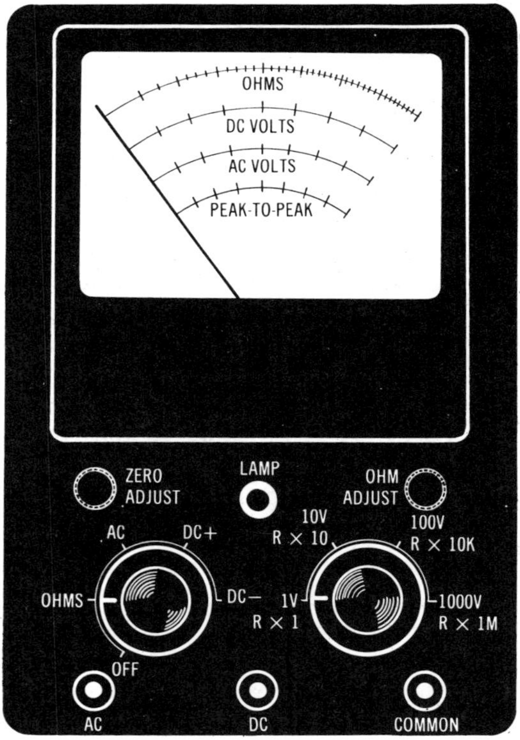

Fig. 2-17. A typical vtvm.

Q2-10. If R_{20} in the circuit in Fig. 2-16 is 100 ohms, what is the value of R_{21}?

Your Answer Should Be:

A2-10. The sum of R_{20} and R_{21} must be 100 times 100 Ω, or 10,000 Ω. Therefore, R_{21} is 10,000 − 100, or **9900 Ω.**

SPECIAL VTVM PRECAUTIONS

1. Beware of high voltages.
2. When making high-voltage checks, grip the probes well up on the insulated parts of the handles. This reduces the danger of shock and the possibility of adding hand capacitance to the circuit.
3. Always ground the ac test probe after a voltage check. The capacitor in the input circuit may have been charged by a dc voltage.
4. An rf probe can be used to measure voltages having frequencies from about 1 kHz and 100 MHz. Connect the probe and its ground connection as close together in the circuit as possible.

WHAT YOU HAVE LEARNED

1. A vacuum-tube voltmeter is similar to a multimeter in function, range selection, and use.
2. A vtvm has a higher input impedance (around 10 megohms) than a multimeter. As a result, it has less of a loading effect on circuits.
3. The range circuits of a vtvm are usually voltage-divider networks. The range switch connects the selected multiplier tap to a vacuum-tube circuit. This circuit regulates the amount of current that will flow through the meter coil.
4. Most vtvm's use some type of bridge circuit. When no measurement is being taken, voltages across the bridge are balanced, and no current flows through the meter. When a measurement is being made, current in one of the tubes increases. This unbalances the voltages across the bridge causing meter current to flow, in proportion to the measurement taken.
5. Safety precautions must be followed when measuring high voltages with a vtvm.

3

Solid-State Multimeters

what you will learn

In this chapter, you will learn the uses of solid-state analog and digital multimeters. You will examine the differences between analog and digital instruments, and learn their advantages and disadvantages. Also, you will learn the methods they use in recording data and the problems encountered in using either type of instrument.

Solid-state instruments have all but replaced vacuum-tube-type instruments in the measurement of voltage, current, and resistance. Solid-state meters, both analog and digital, offer reduced loading of the circuit under test, the same as a vtvm does. Typically, solid-state vom's have a 10- to 15-megohm input impedance. However, solid-state meters have an advantage over the vtvm of being lighter in weight and more portable. Due to their low current drain, solid-state instruments can be operated from batteries over a long period of time. Also, the batteries can be recharged periodically and, thus, the instrument can be operated indefinitely. Solid-state instruments can also be operated from an internal power-supply source just the same as vtvm's. In contrast to the vtvm with its filament-heated cathode, they require no warm-up time.

ANALOG INSTRUMENTS

First, we will consider the analog type of solid-state instrument. That is, an instrument that has one or more calibrated scales and, on which, the measured values are indicated along a scale by a pointer. It is necessary for the user of an analog instrument to interpret the scale reading, and this introduces the possibility of an error.

Solid-state analog instruments are known as solid-state vom's, transistorized vom's, tvom's, and so on. Some instruments use discrete transistors (usually FETs) for the input stage, and some use integrated circuits (ICs) to provide the amplification necessary for the quantity being measured.

The simplified circuit of Fig. 3-1 shows how a basic FET vom works. The multiplier resistors have a total value of 10 megohms. This 10 megohms is the loading, on the circuit being tested, for all ranges of dc voltage. The operator uses the **selector switch** to choose the range that will be within the limitations of the amplifier input circuit. For the circuit shown, you can see that on the 1-volt range, the voltage being measured is applied directly to the FET input circuit. So for this circuit, 1 volt is the highest voltage that can be applied to the FET amplifier. For higher voltages, other ranges will be used. For example, if the voltage being measured is known to be around 85 volts, you would use the 100-V range. This would reduce the FET input voltage to 0.85 volt, which is within the 1 volt maximum input voltage.

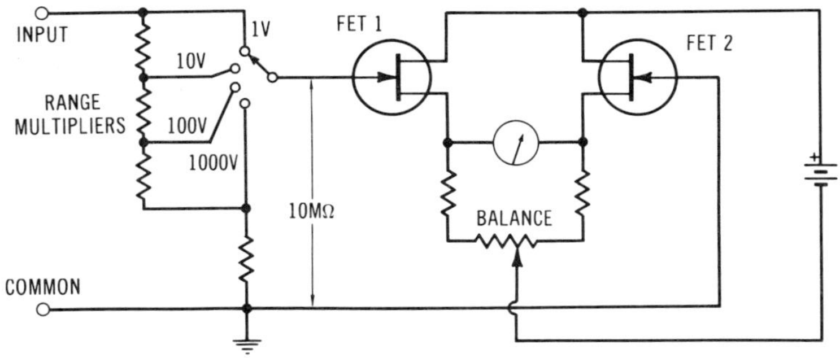

Fig. 3-1. Simplified FET vom circuit.

The two FETs in Fig. 3-1 are selected by the manufacturer as a **balanced pair** and they operate as an **operational amplifier.** This type of amplifier tends to cancel out the effect of any changes in dc power-supply voltage or temperature since the changes will be applied equally to both sides of the meter movement, resulting in a zero current change through the movement.

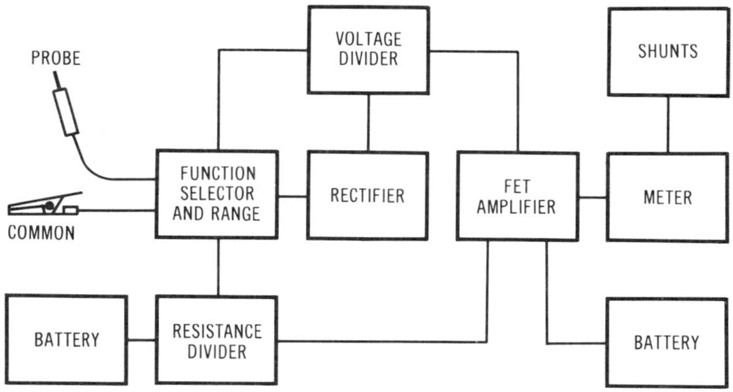

Fig. 3-2. Block diagram of an FET vom.

The block diagram of an FET vom is shown in Fig. 3-2. The function switch selects the function required: ac or dc voltage, ac or dc current, or resistance. For dc voltage measurement, the probe and common lead are connected across the voltage points to be measured. The probe includes an isolating resistor to reduce the loading effect. The function switch connects the voltage being measured to a position on the voltage divider that is suitable for the value being measured. If an ac voltage is being measured, the path goes first through a rectifier and, then, to a voltage divider. After the voltage divider comes an FET amplifier, whose output voltage is applied to the meter. Fig. 3-2 shows that the instrument has two battery power supplies. One supply is for the resistance-measuring circuit. The other operates the solid-state amplifier stages.

In most solid-state vom's, FETs are used rather than bipolar transistors because the FET has a very high input impedance and causes very little loading of the circuit under test.

Q3-1. To provide amplification of the quantity being measured, some solid-state vom's use ____ and others use ____________/__________ as the amplifying device.

Q3-2. Due to the low current drain of solid-state vom's, they can be operated from __________.

Q3-3. The two FETs shown in Fig. 3-1 operate as a ____________ __________.

Your Answers Should Be:

A3-1. To provide amplification of the quantity being measured, some solid-state vom's use **FETs** and others use **integrated circuits** as the amplifying device.

A3-2. Due to the low current drain of solid-state vom's, they can be operated from **batteries.**

A3-3. The two FETs shown in Fig. 3-1 operate as **a differential amplifier.**

DIGITAL MULTIMETERS

Digital instruments have become increasingly popular in every area of electronics. One type of digital instrument that is very widely used is the digital multimeter or dmm. Other versions of the dmm are the digital voltmeter, the digital volt-ohm-milliammeter, and so on.

These digital instruments do not have a calibrated scale and pointer as do the analog instruments. Instead, the measured voltage, current, or resistance value is indicated directly on the front panel of the instrument in the form of a numerical readout or display. The numbers are formed by using LEDs (light-emitting diodes) or LCDs (liquid-crystal displays).

LED Displays

When LEDs are used as the readout, they are usually arranged in the form of a 7-segment display. This is the basic format from which any number can be obtained. The 7-segment display arrangement is shown in Fig. 3-3A, where the segments are identified by the letters a, b, c, d, e, f, and g. By illuminating various elements of the seven LED segments, each of the numerical digits (0 through 9) can be formed. For example, to display a zero, all the segments except segment g are illuminated. This is illustrated in Fig. 3-3B.

An LED, or light-emitting diode, is a semiconductor diode that takes electrical energy and converts it to visible radiation or light. The light may be red, green, or yellow, depending on the semiconductor material and the doping of the material. The color usually used for dmm displays is red.

(A) Segment arrangement.

(B) Use of segments.

Fig. 3-3. Elements of a 7-segment display.

LCD Displays

The basic composition of a LCD, or liquid-crystal display, is two sheets of glass separated by a liquid-crystal material that is sealed in between the glass sheets. A transparent conductive coating is on the outer surface of each of the glass sheets. Electrical leads are connected to character-forming segments on the viewing side of the glass. Voltages applied to the leads will darken the normally transparent material sufficiently to cause numbers or letters to be formed by the darkened areas. No light is generated in this LCD process of forming numbers. Normally, LCD dmm's will have limited usefulness in low light levels.

Q3-4. In the block diagram shown in Fig. 3-2, one battery operates the amplifier stages and the other battery is for the ___________ __________ circuit.

Q3-5. In digital meters, the numerals are formed either by ____ or ____ .

Q3-6. The liquid-crystal display is formed from two sheets of _____ separated by a liquid-crystal material.

> **Your Answers Should Be:**
>
> **A3-4.** In the block diagram shown in Fig. 3-2, one battery operates the amplifier stages and the other is for the **resistance-measuring** stages.
>
> **A3-5.** In digital meters, the numerals are formed either by **LEDs** or **LCDs.**
>
> **A3-6.** The liquid-crystal display is formed from two sheets of **glass** separated by a liquid-crystal material.

Summary

The early dmm's were bench-type instruments. These were followed later by hand-held types which were very convenient to use. The physical arrangement of one type of instrument is shown in Fig. 3-4. Other type models use push-button-type switches for changing the range and function selections rather than the kind of switches shown by this example.

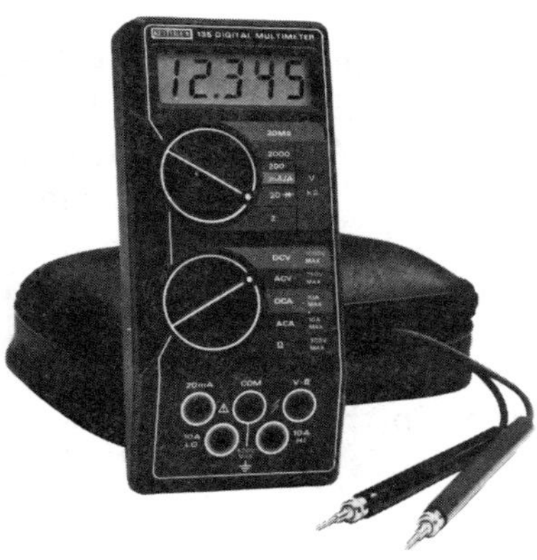

Fig. 3-4. Example of a digital multimeter. (*Courtesy Keithley Instruments, Inc.*)

DMM CONFIGURATIONS

The basic arrangement of a dmm is shown in the block diagram of Fig. 3-5. It includes a signal conditioner, an analog-to-digital (ADC) converter, and a display section. The first stage, the **signal conditioner,** contains the function switch that

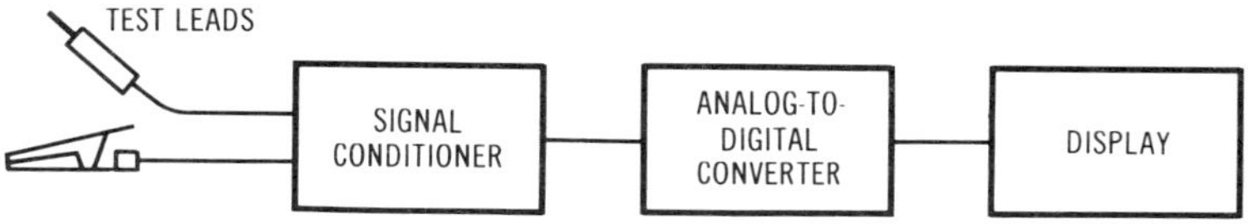

Fig. 3-5. Block diagram of a dmm.

selects both the type of measurement (ohms, positive or negative dc, ac, or current) and the range. Some dmm's have automatic polarity and automatic range selection. The signal conditioner also includes the rectifier for changing an ac voltage to a dc voltage before it goes to the ADC.

The ADC changes the analog signal to a digital signal—that is, to a signal that is represented by a series of pulses. There are several types of ADC circuits used, including linear single-slope, dual-slope, staircase-slope, and integrating types. The single-slope and the dual-slope types are the most often used. As a simplified explanation, a selected interval of the analog voltage to be measured is used to gate an accurately controlled oscillator in the dmm that produces output pulses. These pulses are fed from the oscillator to a digital counter and the counter then displays a digital value that is equal to the voltage being measured.

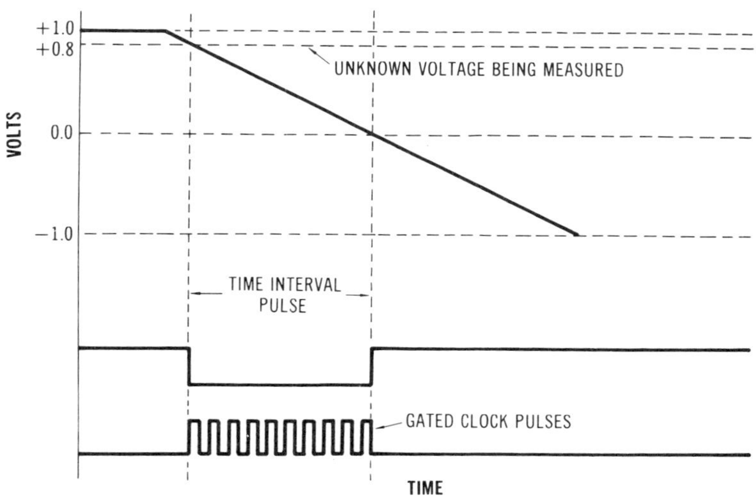

Fig. 3-6. Single-slope ADC waveforms.

Slope Methods of ADC

The single-slope or linear single-slope type of ADC is shown in Fig. 3-6. The slope or ramp extends from +1 volt through 0 volt to −1 volt during a precisely measured time interval. The voltage to be measured is shown as 0.8 volt. When the ramp voltage reaches 0.8 volt (the voltage being measured), a gating pulse starts and the clock pulse generator is started or gated on. Then, when the ramp voltage reaches zero, the clock generator is turned off. The counter counts the pulses. In our example in Fig. 3-6, there are 10 pulses that were generated during the gating-pulse interval. The number of pulses is in direct proportion to the voltage being measured. In our example, 10 pulses represents 0.8 volt so 0.8 volt is displayed on the readout.

Dual-Slope ADC

The main disadvantage of single-slope ADC is its susceptibility to noise. A more widely used method than the single-slope ADC is the dual-slope method shown in Fig. 3-7. With the dual-slope method, the signal to be measured charges a capacitor in the input circuit. The charging period is fixed,

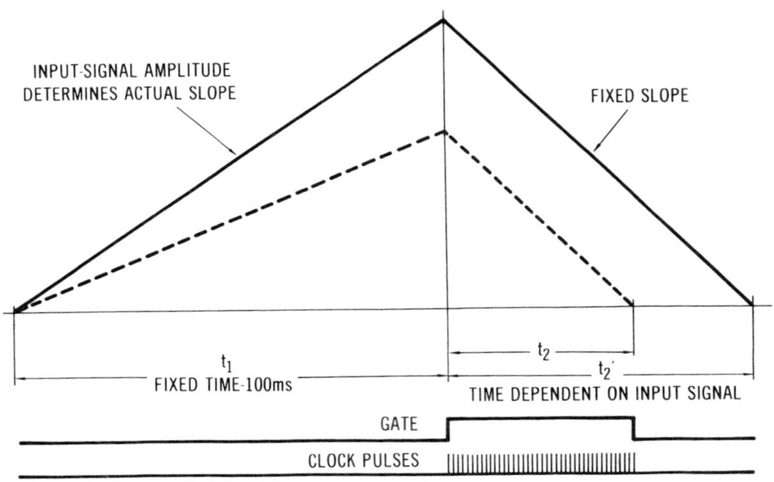

Fig. 3-7. Dual-slope ADC waveforms.

so the amount of charge is proportional to the value of the signal being measured. At the end of the fixed charge period (100 milliseconds, in our example), the charge is dissipated at a constant rate (indicated by the fixed slope in Fig. 3-7) for a time period of t_2 which is set by the value of the input signal. During the period t_2, a gate signal activates a clock pulse oscillator. The number of pulses from the clock is converted by the display circuit into a readout which shows the value of the input signal.

Integrating Method of Conversion

The integrating type of ADC uses the average of the input voltage. This is measured over a predetermined interval. The input to the ADC is an integrating circuit from which the converter gets its name. The method includes an amplifier, shown in Fig. 3-8, that is part of the feedback system. The average value of the measured signal controls the pulse repetition rate of a pulse generator or clock oscillator. Thus, the frequency of the pulses generated is equal to the dc input voltage. A frequency counter indicates in digital form the corresponding value of the voltage measured.

Staircase-Ramp ADC

The staircase-ramp ADC, shown in Fig. 3-9, compares a sample of the input voltage to a staircase ramp generated by the ADC circuit. At a point when the input signal and the staircase ramp are equal, a comparator circuit stops the staircase generator and the digital readout shows the number of

Q3-7. The first stage in a dmm is the ______ ____________.

Q3-8. A digital signal is a signal that is represented by a series of _______.

Q3-9. The two most used methods of analog-to-digital conversion are the _______ ______ and the ____ ______ methods.

Q3-10. In the single-slope method of ADC, when the ramp voltage reaches the input voltage in value, a ______ pulse starts the clock pulse generator.

Q3-11. The clock pulse generator stops when the ramp pulse value reaches ____.

Your Answers Should Be:

A3-7. The first stage in a dmm is the **signal conditioner.**

A3-8. A digital signal is a signal that is represented by a series of **pulses.**

A3-9. The two most used methods of analog-to-digital conversion are the **single-slope** and the **dual-slope** methods.

A3-10. In the single-slope method of ADC, when the ramp voltage reaches the input voltage in value, a **gating** pulse starts the clock pulse generator.

A3-11. The clock pulse generator stops when the ramp pulse value reaches **zero.**

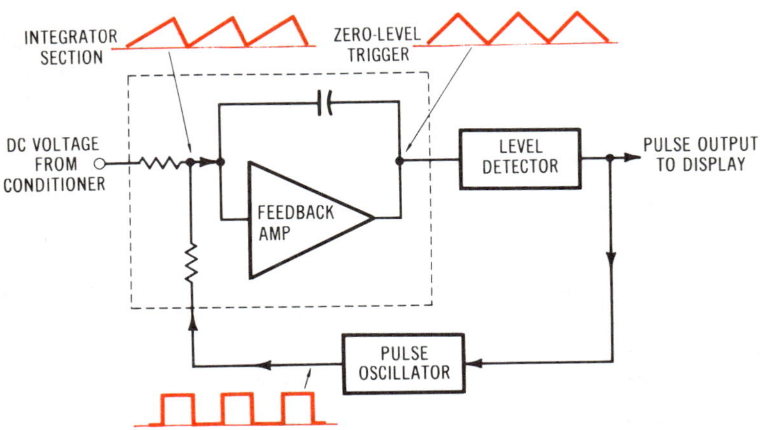

Fig. 3-8. Integrator ADC circuit.

counts used for making the voltages equal. The counts are shown as the voltage value being measured. A 2-Hz sample oscillator refreshes the reading at frequent intervals, typically every one-half second.

DIGITAL OHMMETER CIRCUITS

To measure resistance with a digital ohmmeter, the instrument must provide a source of constant current that flows through the unknown resistance. The dmm measures the volt-

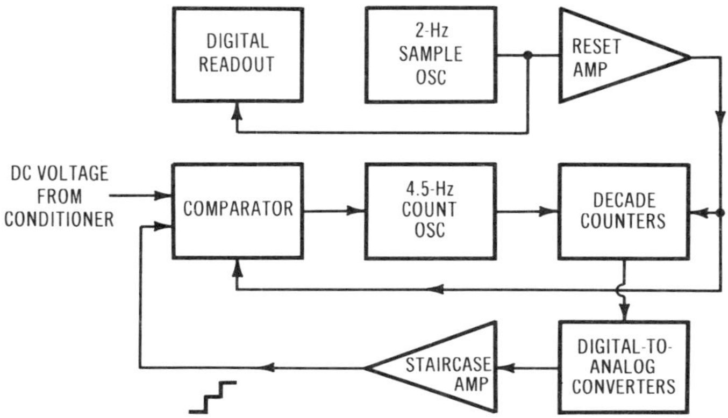

Fig. 3-9. Staircase-ramp ADC circuit. ***(Courtesy Hewlett-Packard Co.)***

tage drop across the resistance and displays that voltage value as the resistance value on the meter. For example, if 1 mA is the constant current and the unknown resistance is 1500 ohms, the voltage across the meter will be 1.5 volts. This will be displayed as 1500 ohms on the digital meter.

In some dmm's, there are high and low ohmmeter ranges. For the low ohms range, the dmm voltage across the resistance is kept to 1 mV or less, in order not to forward bias any semiconductor junctions. Thus, the instrument is then useful for making in-circuit resistance tests in solid-state equipment.

DMM CURRENT MEASUREMENTS

Most of the presently available dmm's now provide for measurement of both ac and dc currents. The ac current-measuring circuit is the same as the dc current-measuring circuit except that, for ac current, the measurement is made by the ac voltage-measuring circuit, which includes a rectifier. In both cases, the measurement is made across an internal shunt resistance. The current being measured flows through this precision-type resistance. The resistance value depends on the current range being used so that the voltage across the resistance is proportional or numerically equal to the current through it. The dmm displays a value equal to the current being measured.

In the example current-measuring circuit shown in Fig. 3-10, assume that the full-range sensitivity of the dmm circuit is 200 mV, so the internal resistance for each current range is then equal to 200 mV divided by the full-range current. For instance, a 200-microampere range would require an internal resistance of 200 mV/ 200 μA, or 1000 ohms.

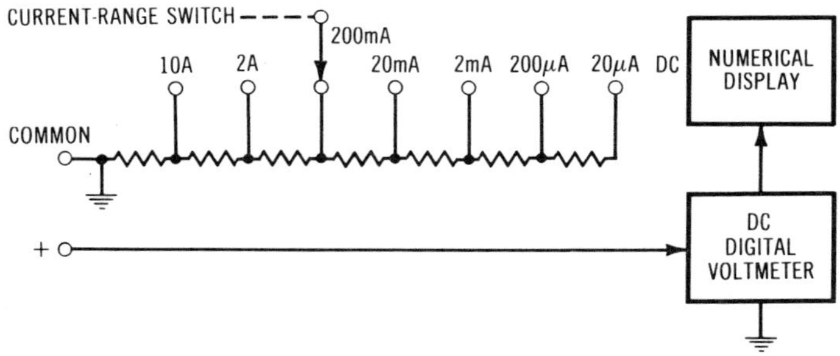

Fig. 3-10. Current-measuring shunts in a dmm.

DMM SPECIFICATIONS

You should become familiar with the more important specifications for digital multimeters, especially those that differ from the specifications for analog instruments. The following discussion will cover some of the dmm specifications.

Ranges

The dmm has different ranges over which it will operate, and these ranges are generally arranged in multiples of 10. A typical dmm might have ranges of 200 mV, 2 V, 20 V, 200 V, and 2000 V. While "2" is the common factor in all of these ranges, the leftmost digit generally will never exceed 1. For example, the highest readings available on the above ranges would be 199 mV, 1.99 V, 19.99 V, 199 V, and 1999 V. A term that you might hear is "overranging." To explain this term, we should say that a manufacturer might consider a 200-mV range to really be a 100-mV range with 100% overranging capability.

Another expression that you will hear is ratings expressed in terms of digits and half digits. For example, an instrument that has a maximum reading of 199 is called a 2½-digit dmm, while if it has a maximum reading of 1999, it's a 3½-digit meter, and if its maximum value is 19999, it's called a 4½-digit dmm. This is illustrated in Fig. 3-11. The half-digit designation is because the leftmost digit is normally limited to 1 instead of 2, such as 1.999 for a 3½-digit dmm rather than 2.000. However, the actual value shown on the meter will depend on the range switch setting and the placement of the decimal point.

Fig. 3-11. Digital displays.

Q3-12. **In the integrating method of ADC, the dc input voltage is equal to the __________ of the pulses generated by the pulse generator.**

Q3-13. **In the staircase ramp ADC, the staircase generator is stopped when the input signal and the __________ _____ are equal.**

Your Answers Should Be:

A3-12. In the integrating method of ADC, the dc input voltage is equal to the **frequency** of the pulses generated by the pulse generator.

A3-13. In the staircase ramp ADC, the staircase generator is stopped when the input signal and the **staircase ramp** are equal.

Some dmm's include automatic ranging (or autoranging) which means that the dmm selects its own range as the measurement is being taken. With such instruments, you need be concerned mainly with having the function selected correctly; the dmm, within limitations, will automatically select the proper range. Some dmm's also include an automatic-polarity selection feature, with a minus sign being placed in the display to indicate if the voltage being measured is negative or opposite in polarity to the way the test leads are connected.

Accuracy

Typically, dmm's have an accuracy rating of 0.1%, plus or minus one least-significant digit. For example, take a voltage of 0.044. A dmm having a rating of 0.1%, plus or minus one least-significant digit, could provide a reading that is in error by as much as 0.00144. This is explained as follows: 0.00044 is 1% of 0.044, and 0.001 is one digit; totaled these values become 0.00144. Thus, the actual voltage value could be anywhere between 0.04256 and 0.04544.

Resolution is a specification closely related to accuracy. A dmm having a 3½-digit display can indicate within 1 mV on a 2-V range; thus, for the 2-V range, the resolution is 0.001.

Resistance Specifications

Many dmm's have 2 megohms as the highest ohmmeter range, but some can measure considerably higher; typically, up to 20 megohms or more. Also, some meters are designed to measure very low ohms, such as continuity between a ground conductor and actual ground. However, when a dmm is connected for a measurement, there is a period of instability until the digit readout stabilizes. This is a disadvantage when making continuity checks. To overcome this disadvantage, the manufacturers of some instruments now use a tone generator to give the user an immediate audible indication when there is good continuity between two points being checked.

Other Features

Most dmm's include features not found in analog meters. These additional features, combined with the way in which a dmm is designed, make the instrument fairly complex. Most instruments include an indicator that tells the user when the instrument is off scale. Typically, an overrange indication is either in the form of flashing lights or in the display of just a special digit.

As far as the type of display is concerned, it is generally either an LED or an LCD. The LED instrument is brighter since the display generates its own light, so LED instruments are more useful in areas of low light levels or darkness. However, the LED sometimes washes out in bright surroundings and out-of-doors. So, where ambient light levels are high, the LCD will be sufficiently visible since its numerals are displayed as dark areas against a light background.

Some dmm's also include automatic polarity selection. Most dmm's usually indicate when the polarity of voltage being measured is negative by including a negative sign in front of the digits displayed. Some dmm's also include a plus sign for positive measurements.

The display rate of a dmm indicates how often the instrument "checks" the measurement being made. Typically, a dmm makes not more than five measurements per second. A display changing faster than that would normally show up only as a flicker.

Q3-14. For resistance measurement in a dmm, a constant ________ source is used to create a voltage drop across the unknown resistance.

Q3-15. The ranges for a dmm are generally specified in multiples of ___.

Q3-16. A manufacturer who lists 100% overranging for a 100-mV range is actually describing a ____-mV range.

Q3-17. A dmm that has a maximum reading of 199 is a ____-digit dmm.

Your Answers Should Be:

A3-14. For resistance measurement in a dmm, a constant **current** source is used to create a voltage drop across the unknown resistance.

A3-15. The ranges for a dmm are generally specified in multiples of **10.**

A3-16. A manufacturer who lists 100% overranging for a 100-mV range is actually describing a **200**-mV range.

A3-17. A dmm that has a maximum reading of 199 is a 2½-digit dmm.

ADVANTAGES AND DISADVANTAGES OF DMMs

The usual dmm is 10 times, or more, accurate than a vom, a vtvm, or a solid-state vom. The dmm has an input resistance range equal to, or better than, the other electronic voltmeters used for dc voltage measurements. With a dmm, there is no such thing as parallax viewing error (due to viewing angle) as there is with analog instruments. A dmm also provides some automatic features not found on analog instruments that include automatic zeroing, automatic decimal locating, polarity indication and/or automatic polarity, and tone or visual continuity indication.

One disadvantage of the dmm is that there is a brief period of instability when it is first connected in a circuit for a measurement, or when a change of voltage, current, or resistance value occurs. Thus, it is awkward to use when checking dc voltages that have pulses, ripple, or varying levels. As a result, the use of a dmm is limited when used for peaking or tuning circuits. The frequency response of a dmm is typically limited to 400 Hz, although some dmm's can measure to 75 kHz or more.

Some problems of the earlier dmm's were a susceptibility to interference from nearby rf fields and from pulses in tv receivers, plus a problem of obtaining unreliable readings when measuring dc voltages in half-wave rectifiers, but these advantages have been eliminated in the later instruments.

WHAT YOU HAVE LEARNED

1. Solid-state devices have all but replaced the vacuum tube in electronic test instruments.
2. Some special advantages of solid-state instruments are that they have practically zero warm-up time, they require little power for operation so they may be run on batteries, and they are lighter in weight and, therefore, more portable.
3. Analog instruments are those with a calibrated scale and a pointer. Analog instruments are still being used although they have been replaced in many cases by digital instruments.
4. A digital multimeter, or dmm, does not have a calibrated scale or pointer, but, instead, displays measured values as discrete numerals, 0 through 9, so that any possible misinterpretation error is eliminated.
5. The numeric displays are made up of seven diode segments which are illuminated in different combinations to produce numerals or other characters.
6. There are mainly two types of digital displays—light-emitting diodes (LEDs), and liquid-crystal displays (LCDs).
7. LEDs are easier seen than LCDs where light levels are low; LCDs are better in high surrounding light.
8. The main sections of a dmm are the signal conditioning section, the analog-to-digital converter (ADC) section, and the display section.
9. The signal conditioner selects the range and function for the quantity being measured, the ADC changes the analog quantity being measured to a digital quantity, and the display section displays the measured value.

Q3-18. The (LED or LCD) type of display is more useful where light levels are high, whereas, the (LED or LCD) display is better in poor light levels.

Q3-19. Parallax viewing error is a type of error that (is, is not) possible with dmm's.

Q3-20. The dmm (is, is not) especially useful for tuning or peaking electronic circuits.

Your Answers Should Be:

A3-18. The **LCD** type of display is more useful where light levels are high, whereas, the **LED** display is better in poor light levels.

A3-19. Parallax viewing error is a type of error that **is not** possible with dmm's.

A3-20. The dmm **is not** especially useful for tuning or peaking electronic circuits.

10. The two main types of a/d converters are the single-slope ADC and the dual-slope ADC, although some more advanced types of instruments use a triple-slope method.
11. Other methods of analog-to-digital conversion are the integrating method and the staircase-ramp method.
12. For resistance measurements, a digital multimeter includes a constant current which is applied to the resistor being measured. The voltage drop across the unknown resistance is then displayed in terms of resistance.
13. Digital multimeter ranges are in multiples of 10, such as 200 mV, 20 mV, 2 mV, 2 V, 20 V, and so on.
14. Dmm's are described in terms of so many digits plus a half digit. A dmm having a maximum display of 199 would be described as a 2½-digit instrument. One having a maximum display of 1999 would be a 3½-digit instrument, and so on.
15. Dmm accuracy is described as a certain percentage of the reading, such as 0.1%, plus or minus one least-significant digit.
16. Some digital multimeters include an audible or a visual indicator that is used for fast checking of continuity because dmm stability, when first connected for a measurement, is usually not fast enough to permit making quick checks.

4

The Oscilloscope

what you will learn

An oscilloscope is a test instrument capable of showing the waveforms of sinusoidal and nonsinusoidal signals. You will learn how the oscilloscope performs its various functions to aid the user in gaining valuable knowledge about an electronic circuit. In this chapter, you will discover how the oscilloscope can be used to measure the voltage and phase of an applied signal. This chapter also gives information about the operation of the cathode-ray tube.

LIMITATIONS OF METERS

You have become acquainted with multimeters, vacuum-tube voltmeters and solid-state vom's. If asked to describe them in a brief statement, you might say they are instruments capable of measuring the magnitude (size) of certain electrical characteristics. This would be a good description if you added that the characteristics are basically limited to voltage, current, and resistance. How much information would a multimeter or vtvm tell you about a voltage that varies as shown in the sketch of Fig. 4-1?

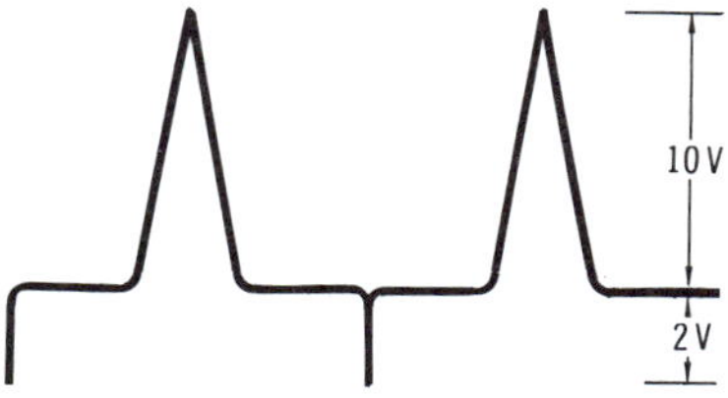

Fig. 4-1. A voltage waveform.

Your answer might be merely, "Voltage." This would be a good answer, since you did not specify the amount of voltage. A multimeter, or a vtvm, is designed and has its scales calibrated to measure sinusoidal (sine-wave) ac voltages. It cannot accurately measure a nonsinusoidal voltage. Most meter pointers will not be able to follow the rise and fall of such a voltage waveform, and only a slight indication will be obtained. However, some vtvm's will read peak-to-peak voltages.

IMPORTANCE OF WAVEFORMS

Since a voltage or current can be described in terms of amplitude and time, you can identify and analyze any signal in these terms. A graph or picture (Fig. 4-2) of how the amplitude of a signal varies with time is called a **waveform.**

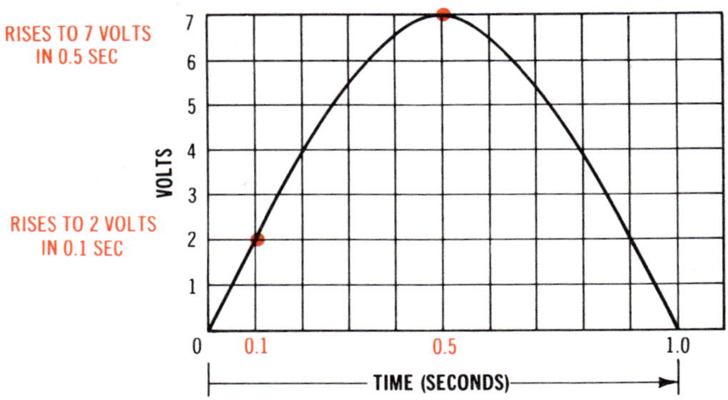

Fig. 4-2. A signal varying in amplitude and time.

To maintain, troubleshoot, and repair electronic equipment, a technician needs to look at the waveform of a signal as it passes from one circuit to another. For this, an instrument is needed that will provide a reliable representation of the signal. If the representation matches the desired size and shape of a signal that should occur at the test point, the technician can assume that the circuit is operating as it should. If the representation does not match the signal, the type and amount of difference will help in identifying the cause of the trouble.

Circuit Characteristics

Each electronic circuit is designed to accomplish a specific purpose. The purpose determines the input and output requirements of the circuit. The input signal of one circuit is normally the output signal of the preceding circuit (or stage). The output signal is the signal required as the input to the next stage. Circuit components are selected and connected in such a way as to convert the input signal to the required output for each

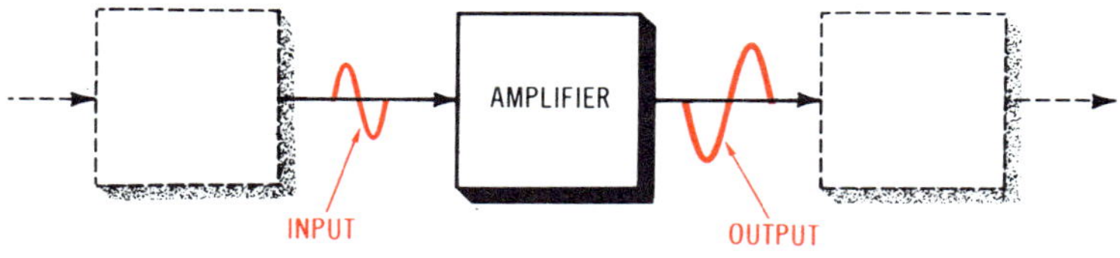

Fig. 4-3. An amplifier stage.

stage. An amplifier, for example, usually receives a small signal from a preceding stage and converts it to a larger signal (Fig. 4-3). In other words, the stage amplifies the signal.

It is often helpful to be able to determine if the change from input to output signal has been made properly. For example, it is desirable to know if the shape of a signal waveform has changed when the signal passes through an amplifier circuit. In Fig. 4-4, the leading half of the cycle in the output has been

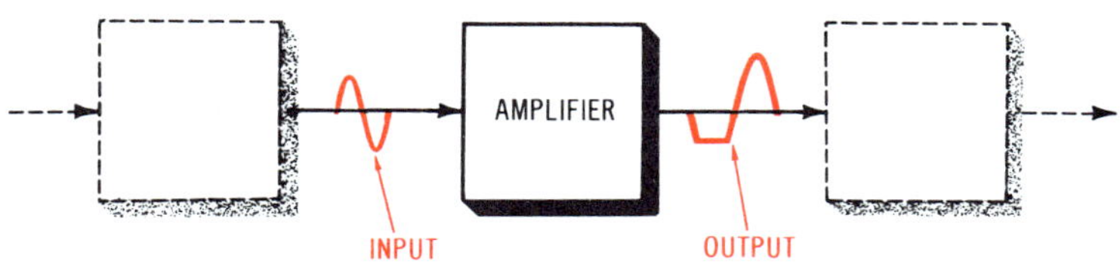

Fig. 4-4. Distorted output.

distorted. You would, therefore, suspect that the amplifier had gone into saturation and was clipping off that portion of the wave. You could also reasonably conclude that the most probable cause of the trouble was a change in tube bias.

Q4-1. What determines the input and output requirements of a circuit?

Q4-2. What signal characteristics are shown by a waveform?

Your Answers Should Be:

A4-1. The input and output requirements of a circuit are determined by the **purpose of the circuit.**

A4-2. The **amplitude and time** of a signal are shown by a waveform.

Waveform Characteristics

Each pulse in Fig. 4-5 has an amplitude of 3.5 volts and a width of 1 millisecond. The pulse repeats itself every 10 milli-

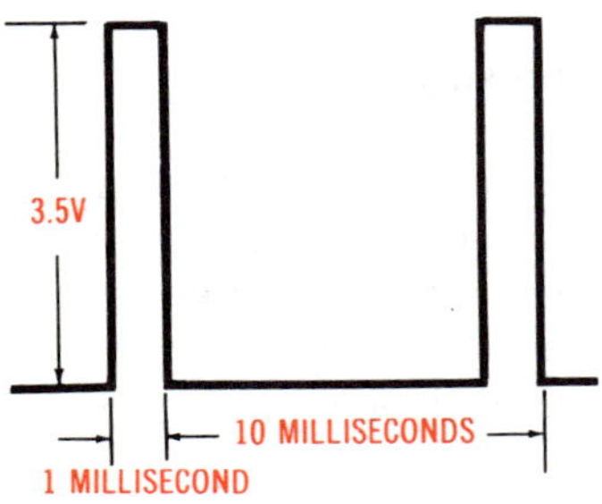

Fig. 4-5. A pulse waveform.

seconds. Since one pulse occurs every 0.01 second, the pulse frequency is 100 pulses per second.

$$\text{Frequency} = \frac{1}{\text{time}}$$

A **sine wave** is a curved waveform. There are other waves

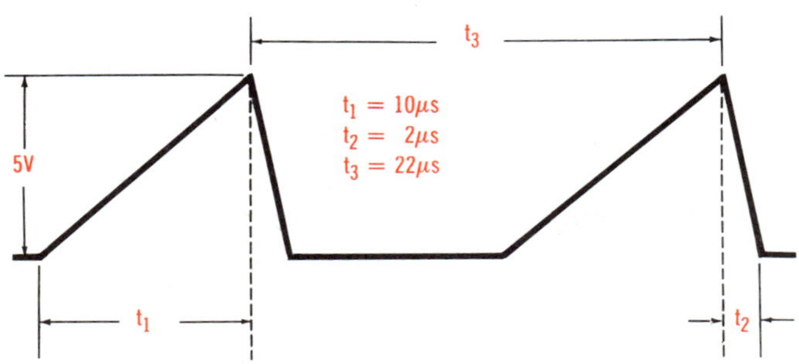

Fig. 4-6. A sawtooth waveform.

whose increases and decreases appear to be straight lines. The pulse waveform in Fig. 4-5 is an example. Another example is a **sawtooth** waveform (Fig. 4-6).

Another observation that can be made about waveforms is their phase relationship. What is the relationship of their amplitudes at a given instant of time? Observe the two waveforms in the illustration of Fig. 4-7.

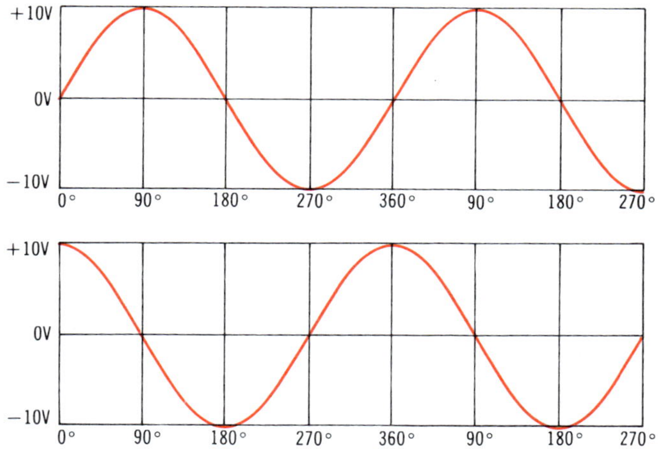

Fig. 4-7. Phase relationship.

The top waveform (sine wave) rises from zero to 10 volts positive in 90° (one-quarter cycle). During the same time period, the bottom sine wave decreases from 10 volts to zero. In other words, the two are 90° out of phase.

These and other characteristics of waveforms can be determined by plotting amplitude against time. Even if he had the means of measuring small changes in time, man's vision is too slow to follow the rapid rise and fall of the amplitude. He could not make an accurate plot. The oscilloscope does this for him electronically. It presents a pictorial representation of an amplitude-versus-time plot of the waveform.

Q4-3. What is one type of waveform that appears to be made up of straight-line segments?

Q4-4. Why does the electronics troubleshooter need an oscilloscope?

Your Answers Should Be:

A4-3. A **sawtooth waveform** is one that appears to be made up of straight-line segments.

A4-4. The electronics troubleshooter needs an oscilloscope **to see an accurate representation of various circuit signals.**

WHAT IS AN OSCILLOSCOPE?

An oscilloscope is an indicator. It indicates the shape of a signal appearing at a test point. Some oscilloscopes are better at showing a reliable reproduction of waveforms than others. The difference is merely one of design. All oscilloscopes function in accordance with the same set of fundamentals. If you learn how one oscilloscope works and how it can be used, you can easily learn how to operate others.

All oscilloscopes contain a **cathode-ray tube** (crt) and a group of **control circuits.** The crt displays the waveform. The control circuits present the signal to the crt. A set of test leads brings the waveform to the control circuits. This is illustrated by Fig. 4-8.

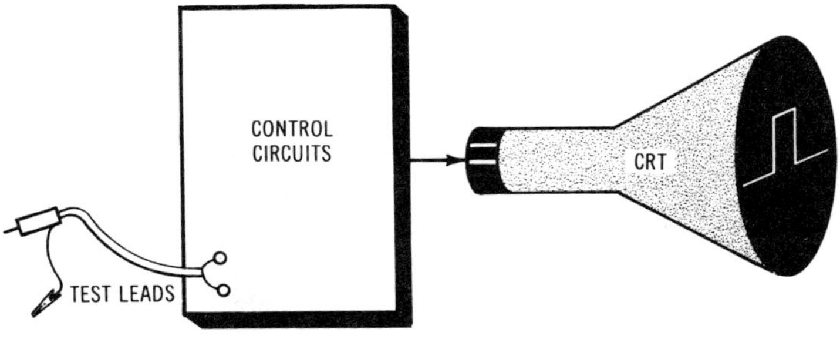

Fig. 4-8. The oscilloscope.

A typical oscilloscope will be described in terms of how the cathode-ray tube and the control circuits function. The test leads are only slightly different from those with which you are already familiar. Since the control circuits are designed to operate the crt, the cathode-ray tube will be studied first.

CATHODE-RAY TUBES

The cathode-ray tube is a vital part of a television set. The crt operates by moving a controllable beam of electrons across the inside face of the tube. The number of electrons in the beam is determined by the blacks, grays, and whites of the scene that the tv camera is viewing. White is produced by a large number of electrons striking a chemical coating on the inside of the tube. The electrons cause the coating to give off light. Black is achieved by stopping the electron flow, and shades of gray are obtained by varying the amount of electrons striking the screen to some quantity that is between the amounts required for black and white.

The picture is "painted" on the screen by the narrow electron beam moving back and forth across the tube many times a second. This movement is due to a varying magnetic field produced by a set of coils around the neck of the crt.

The principle of putting a picture of a waveform on the screen of an oscilloscope is similar. The movement of an electron beam is controlled **electrostatically** so that the beam traces out the pattern of the waveform being measured. As in the tv tube, electrons illuminate a coating on the inside of the tube.

Electrostatics

To understand how a crt operates requires a review of what you learned about **electrostatic fields.** As you recall, an electrostatic field is a region in which electric forces are acting.

An electrostatic field can be developed between two charged plates. If one plate is negative with respect to the other, the direction of the electric force can be determined.

Q4-5. A waveform can be described in terms of its vertical and horizontal dimensions. What are these dimensions?

Q4-6. A cathode-ray tube can display a picture on its face, or screen. What causes the picture to appear?

Q4-7. An oscilloscope is made up of a cathode-ray tube and a group of control circuits. What is the function of the control circuits?

Q4-8. What is an electrostatic field?

Your Answers Should Be:

A4-5. The vertical and horizontal dimensions of a waveform are **amplitude and time.**

A4-6. The picture on a crt is developed by a **moving electron beam** that strikes and illuminates a chemical coating on the inside face of the tube.

A4-7. The function of the oscilloscope control circuits is to **present a signal to the crt.**

A4-8. An electrostatic field is a **region in which electric forces are acting.**

Forces in an Electrostatic Field

In Fig. 4-9, lines of electric force take a direction from negative to positive. This means a negatively charged body entering the field would be moved downward (from negative to positive). A positively charged body, however, would be moved upward (positive to negative). Like charges repel, and unlike charges attract. Do you recall how an electrostatic field is formed?

An electrostatic field is formed with a voltage source and a pair of metallic plates to hold the charges.

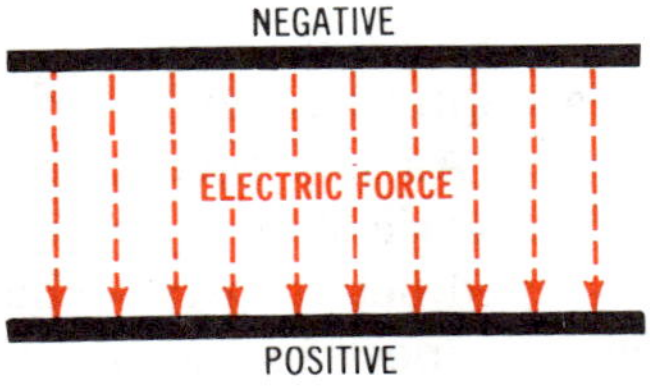

Fig. 4-9. Lines of electric force.

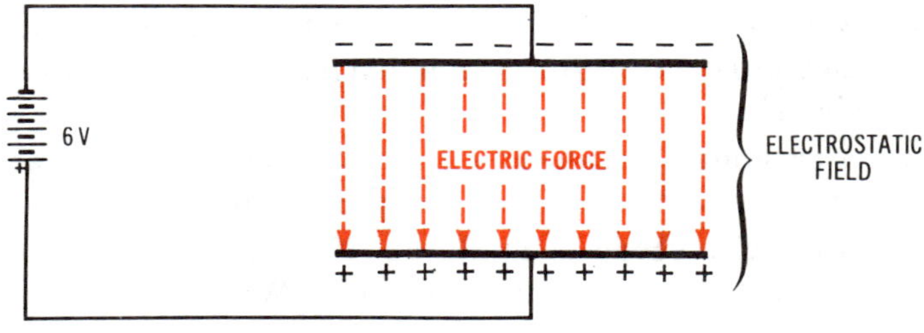

Fig. 4-10. An electrostatic field.

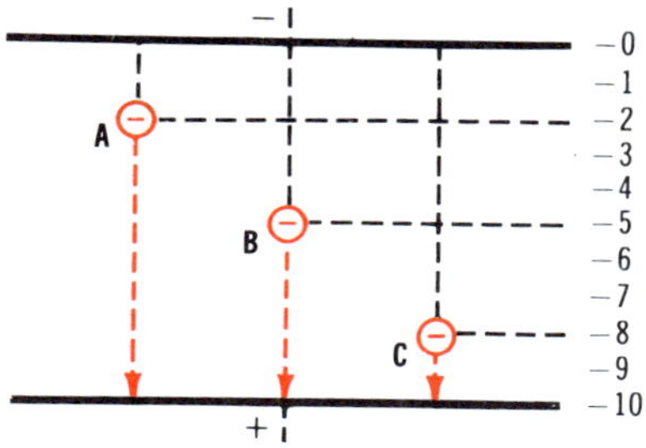

Fig. 4-11. A uniform electric field.

If a 6-volt battery is connected to the plates in the manner shown in Fig. 4-10, the battery will draw electrons from the bottom plate and deposit them on the top plate until the difference in potential between the plates equals the battery voltage. The potential of the plate having an excess of electrons will be negative. The other plate, being deficient in electrons, will be positive.

As indicated in the diagram of Fig. 4-10, an electric force exists in an electrostatic field. This force can act on other charges entering the field.

In Fig. 4-11, three electrons are located in an electrostatic field. All three are attracted by the positive plate and repelled by the negative plate. The distance between the plates is marked off in 10 equal units; electron A is 2 units away from the negative plate, electron B is 5 units away, and electron C is 8 units removed.

A uniform electric field is established between the two plates. This means that an electron in one part of the field has the same force acting on it as an electron in another part of the field. Thus, electrons A, B, and C all have the same amount of force acting on them, even though each is in a different position relative to the plates. Since the same amount of force is exerted on each electron, the relative time of travel of each electron depends on its distance from the positive plate.

Q4-9. A positive ion rests in an electrostatic field. Toward which plate will it move?

Q4-10. What causes an electrostatic field to exist between two metallic plates?

Your Answers Should Be:

A4-9. The positive ion will go **toward the negative plate.**

A4-10. An electrostatic field is formed when **one plate has an excess of electrons and the other plate has a deficiency of electrons.**

Distribution of Electric Force

Electrostatic force is conventionally represented by dashed lines with an arrowhead showing the direction in which the force is acting. Is an individual force represented by each line? If this were true, an electron traveling between the plates with sufficient velocity to pass through the field would cover the distance in a stair-step pattern, as shown in Fig. 4-12A. However, it is known that the path of an electron through a force field is curved, as shown in Fig. 4-12B. Force does not exist in distinct beams; it is continuous and uniform across the field. However, it is easier to talk about the field in terms of imaginary lines.

Distribution of Force Lines

Thus far, you have visualized lines of electric force as being straight and parallel to each other. This is not always true. The diagram shown in Fig. 4-13 illustrates that part of the lines of force can take a curved direction. Remember, the lines shown are only representative of a continuous field.

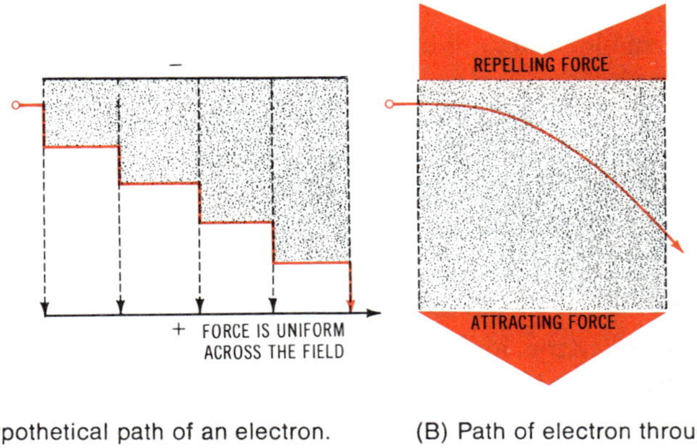

(A) Hypothetical path of an electron. (B) Path of electron through field.

Fig. 4-12. Distribution of electric force.

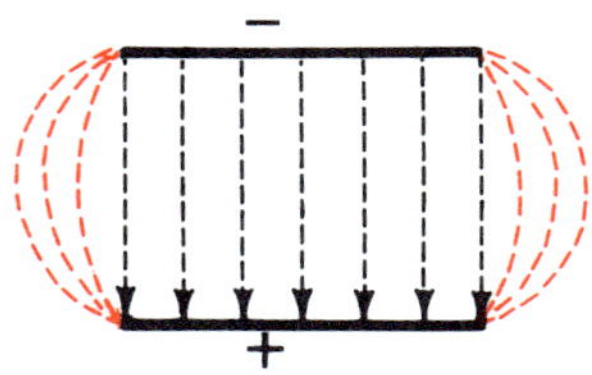

Fig. 4-13. Distribution of force lines.

Lines directly between the plates are parallel to each other. Because they are equal lines of force, they tend to repel each other in a horizontal direction. The repelling effect is equal in all directions around a line of force. The lines of force at the edges of the field are bent outward because there are no lines outside the field to repel them inward.

An electron in motion through an electric field will tend to follow the direction of the lines of force. The amount that the electron path will bend in the direction of the lines of force depends on the velocity of the electron and the potential of the electric field (see Fig. 4-14). A fast electron may speed through the field with little curvature to its path. The path of a slow electron would curve more. Electrons of equal velocity would curve more when they are passing through a strong rather than a weak field.

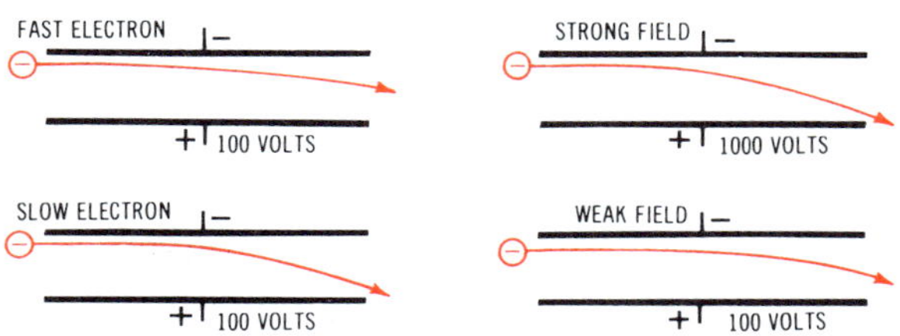

(A) Same field strength. (B) Same electron speed.

Fig. 4-14. Velocity of electron and potential of electric field.

Q4-11. Does an electrostatic field consist of distinct and separate lines of force or is it continuous?

Q4-12. Does a moving electron tend to take a path parallel to the direction of force or perpendicular to it?

Q4-13. Why are the lines of force at the edge of an electrostatic field curved?

> **Your Answers Should Be:**
>
> **A4-11.** An electrostatic field is **continuous.**
>
> **A4-12.** A moving electron will **tend to** take a path **parallel** to the direction of electric force.
>
> **A4-13.** The lines of force at the edge of an electrostatic field are curved because **there are no lines of force outside the field to repel them inward.**

Electrostatic Forces Between Circular and Tubular Plates

In the diagram of Fig. 4-15, an electrostatic field is shown between two plates having center holes. Observe the curvature of the force lines under the holes.

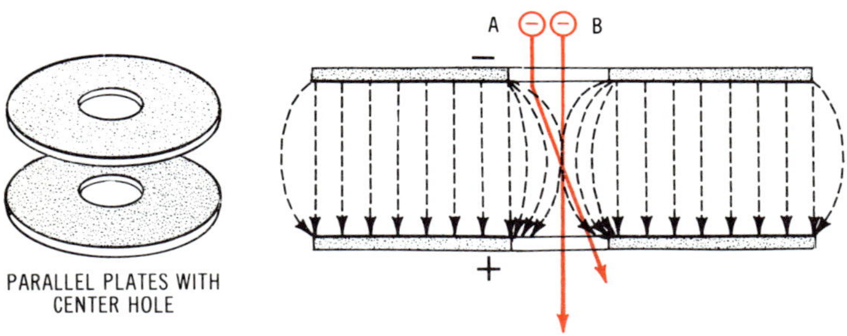

Fig. 4-15. Forces between circular plates.

Since its path is parallel to the force lines, electron B will pass straight through the axis (center line) of the holes. Electron A starts in the same direction as electron B. However, when electron A enters the field, it turns in the direction of the force lines. Just before it leaves the field, it is turned even further and travels in the direction of the curvature of the force lines.

Suppose a small and a large cylinder, both charged with a positive potential, are placed so the electrons must pass through them. Also, suppose the larger cylinder has a more positive charge. The distribution of the lines of force would look like the illustration in Fig. 4-16.

An electron in the space at the left of the small cylinder will be attracted toward the cylinder by the positive charge. If the electron is traveling along the axis of the cylinder, it will pass through without crossing a line of force. As it approaches the larger, more positively charged cylinder, the velocity of the electron will increase.

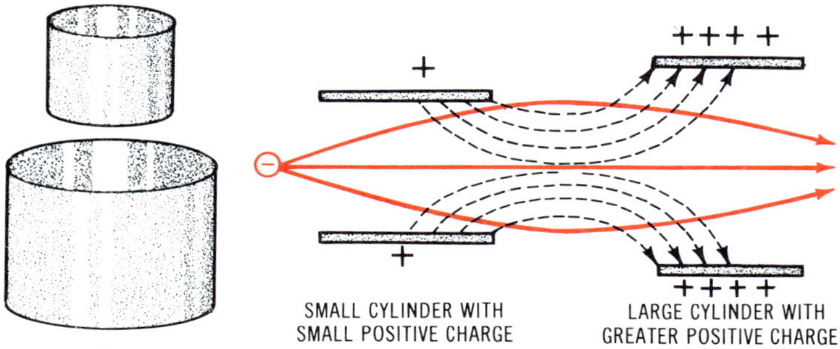

Fig. 4-16. Forces between tubular plates.

An electron entering the small cylinder at an angle will cut the lines of force and be turned in their direction as shown by the top and bottom electron paths in Fig. 4-16. As it approaches the larger cylinder, the electron will be accelerated by the higher positive potential. Because of the higher electron velocity, the force lines in the larger cylinder will have a smaller turning effect on the electron. If the difference of potential between the cylinders is adjusted properly, the electrons will unite at a given distance after passing through the second cylinder. This action of the electrons as they pass through the influence of the two cylinders provides a convenient method of **focusing** the electron beam.

Q4-14. As an electron approaches the larger cylinder, the velocity of the electron will _ _ _ _ _ _ _ _ _ .

Q4-15. Why is the above statement true?

Q4-16. What path will the electron take if it is on the axis of both cylinders as it enters the first?

Q4-17. What path will the electron take if the small cylinder is charged positively and the large cylinder is charged negatively?

Your Answers Should Be:

A4-14. As an electron approaches the larger cylinder, the velocity of the electron will **increase.**

A4-15. The above statement is true because **the larger cylinder is more positively charged.** It will attract the electron with a greater force, thereby increasing the velocity of the electron.

A4-16. The electron will move in a **straight path** through both cylinders.

A4-17. The electron will be **attracted by the small cylinder,** but **repelled by the large cylinder.**

ELECTRON GUN

Cathode-ray tubes (Fig. 4-17) used in oscilloscopes consist of an **electron gun, a deflection system,** and a **fluorescent screen.** All elements are enclosed in an evacuated container, usually glass. The electron gun generates electrons and focuses them into a narrow beam. The deflection system moves the beam across the screen in the manner desired. The screen is coated with a material that glows when struck by the electrons.

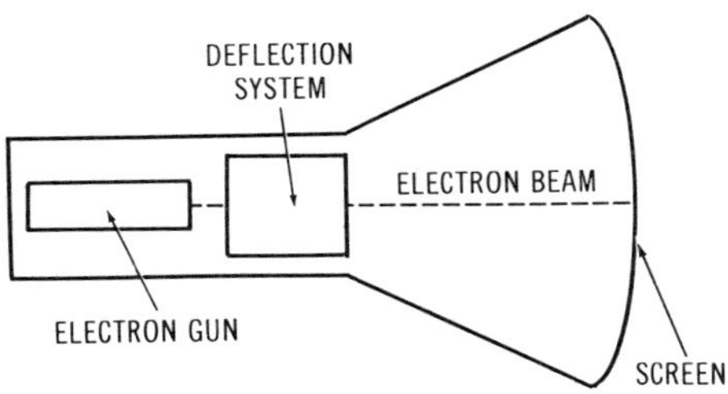

Fig. 4-17. Cathode-ray tube.

An electron gun (Fig. 4-18) has a cathode to generate electrons, a grid to control electron flow, and a positive element to accelerate electron movement. The control grid is cylindrical in shape and has a small opening in a **baffle** located at one end. The positive element consists of two cylinders, called **anodes.** They also contain baffles (or plates) having small holes in their centers. The main purpose of the first anode is to focus the electrons into a narrow beam on the screen. The second anode speeds up the electrons as they pass.

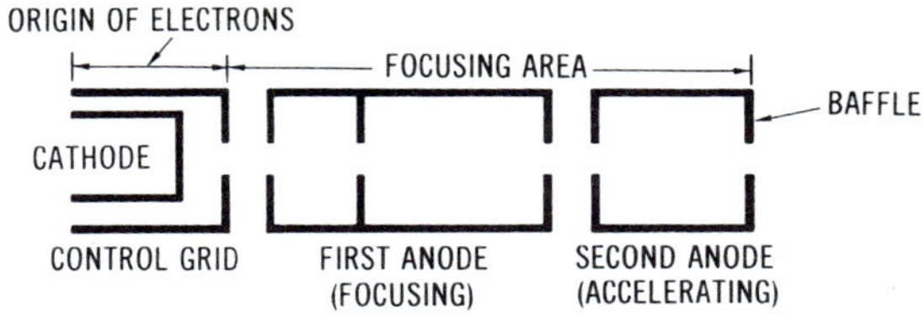

Fig. 4-18. Diagram of an electron gun.

Cathode and Grid

The cathode (Fig. 4-19) is indirectly heated and emits a cloud of electrons. The control grid is a hollow metal tube placed over the cathode. A small opening is located in the center of a baffle at the end opposite the cathode. The grid is maintained at a negative potential with respect to the cathode.

A high positive potential on the anodes (Fig. 4-19) pulls electrons through the hole in the grid. Since the grid is near the cathode, it can control the number of electrons that are emitted. As in an ordinary vacuum tube, the negative voltage of the grid can be changed to vary electron flow or stop it completely. The brightness of the image on the fluorescent screen is determined by the number of electrons striking the screen. **Intensity** (brightness) can, therefore, be controlled by the voltage on the control grid.

Focus Control

Focusing is accomplished by controlling the electrostatic fields that exist between the grid and first anode and between the first and second anodes. Study the diagram in Fig. 4-19. See if you can show the paths of electrons through the gun.

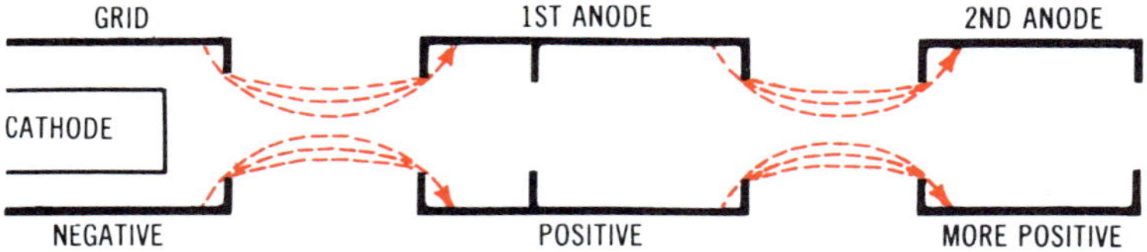

Fig. 4-19. Electrostatic fields.

Q4-18. Which element controls the number of electrons striking the screen?

Q4-19. Which element controls the focus of the beam?

> **Your Answers Should Be:**
>
> **A4-18.** The **control grid** controls the number of electrons striking the screen.
>
> **A4-19.** The **first anode** controls the focus of the beam.

Electrostatic Lenses

The diagram in Fig. 4-20 illustrates electrons moving through the gun. The electrostatic field areas are often referred to as **lenses.** The first electrostatic lens causes the electrons to cross at a focal point within the field. The second lens bends the spreading streams and returns them to a new focal point.

The diagram also shows the voltage relationships on the electron-gun elements. The cathode is at a fixed positive voltage with respect to ground. The grid is at a variable negative voltage with respect to the cathode. A fixed positive voltage of several thousand volts is connected to the second (accelerating) anode. The potential of the first (focusing) anode is less positive than the potential of the second anode. It can be varied to place the focal point of the electron beam on the screen of the tube. Control-grid potential is established at the proper level to allow the correct number of electrons through the gun for the desired screen intensity.

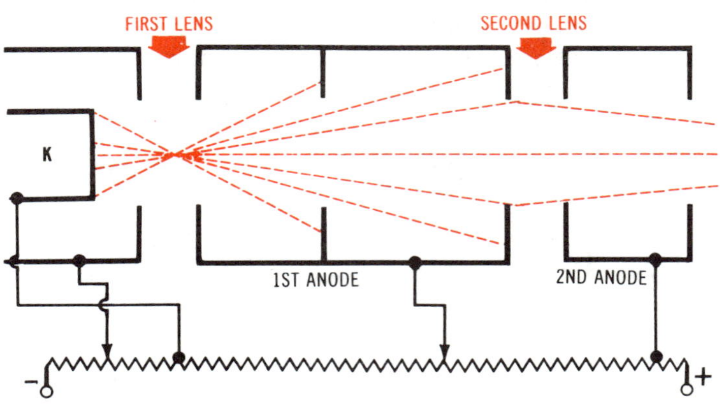

Fig. 4-20. Formation of electron beam.

ELECTRON-BEAM DEFLECTION SYSTEM

The electron beam is developed, focused, and accelerated by the electron gun. It appears on the screen of the crt as a small bright dot (Fig. 4-21). If the beam is left in one position, the electrons will soon burn away the illuminating coating in that one area. To be of any use, the beam must move. As you have learned, an electrostatic field can bend the path of a moving electron.

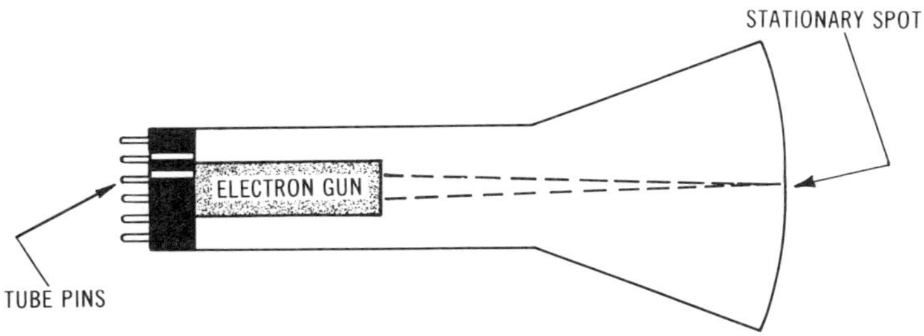

Fig. 4-21. The electron beam.

Assume the beam of electrons passes through an electrostatic field between two plates. Since electrons are negatively charged, they will be deflected in the direction of the electric force (from negative to positive). The electrons will follow a curved path through the field. When the electrons leave the field, they will take a straight path to the screen at the angle at which they left the field. Although the beam is still wide (the focal point is at the screen), all the electrons will be traveling toward the same spot. This is assuming, of course, that the proper voltages are existing on the anodes which produce the electrostatic field. Changing the voltages changes the focal point of the beam.

Factors Influencing Deflection

The **angle of deflection** (the angle the outgoing electron beam makes with the axis between the plates) depends on several factors (Fig. 4-22). These factors include the length of

Q4-20. Why are the electrostatic fields between electron-gun elements called lenses?

Q4-21. What is the function of the second anode?

Your Answers Should Be:

A4-20. They are called lenses because **the fields bend electron streams** in the same manner that optical lenses bend light rays.

A4-21. The second anode **accelerates the electrons** emerging from the first anode.

the deflection field, spacing between the deflection plates, difference of potential between the plates, and accelerating voltage on the second anode.

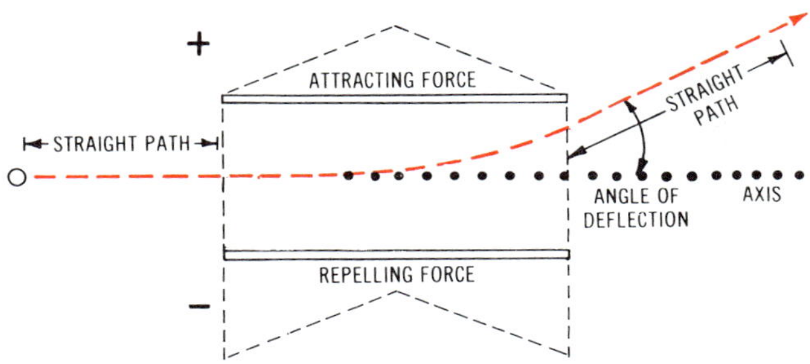

Fig. 4-22. Angle of deflection.

Length of Field—A long field has more time to exert its deflecting forces on an electron beam than has a shorter field (Fig. 4-23). Therefore, it bends the beam to a greater deflection angle. This fact assumes that all other factors are equal.

Spacing Between Plates—The closer the plates, the more effect the electric force has on the electron beam (Fig. 4-24).

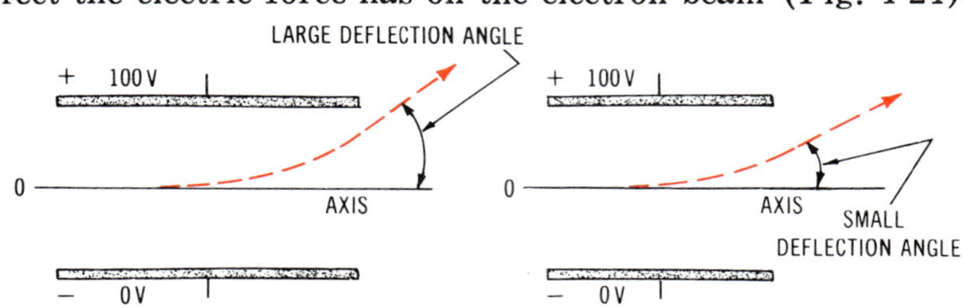

(A) Long deflection plates.

(B) Short deflection plates.

Fig. 4-23. Length of field.

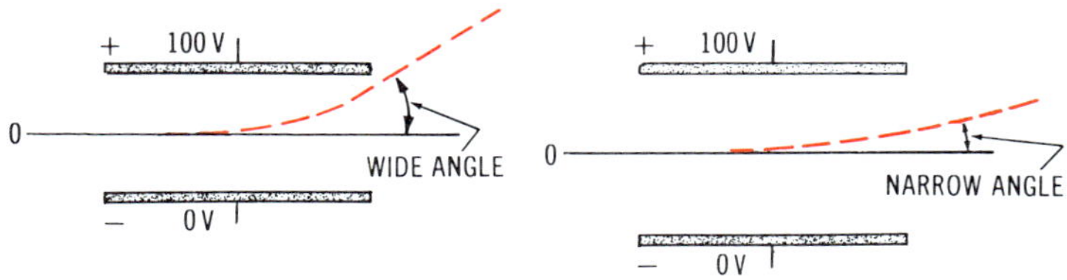

(A) Close spacing. (B) Wide spacing.

Fig. 4-24. Spacing between the plates.

Difference of Potential—Intensity of the electric force can be varied by the plates' difference of potential (Fig. 4-25).

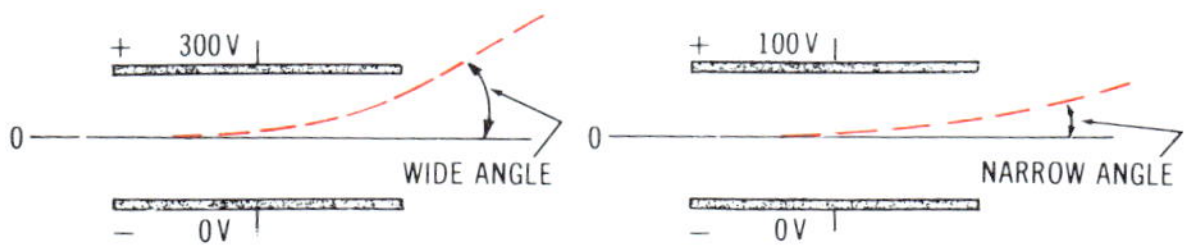

(A) High potential. (B) Low potential.

Fig. 4-25. A difference of potential.

Beam Acceleration—The faster the electrons are moving, the smaller their deflection angle will be (Fig. 4-26).

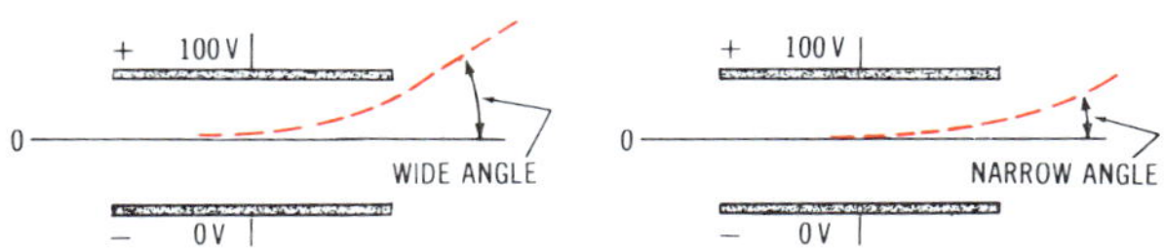

(A) Low acceleration. (B) High acceleration.

Fig. 4-26. Beam acceleration.

Q4-22. Is the deflection angle in a crt more easily changed by plate spacing or plate potential?

Q4-23. The deflection angle is more with (long, short) plates.

Q4-24. Is the deflection angle greater with close or wide spacing of plates?

Q4-25. Is the deflection angle greater with a high or a low potential on the plates?

Q4-26. Is the deflection angle greater when the beam is moving fast or when it is moving slow?

Your Answers Should Be:

A4-22. **Plate potential** is more convenient to control.

A4-23. The deflection angle is more with **long** plates.

A4-24. It is greater with **close spacing.**

A4-25. It is greater with **a high potential.**

A4-26. It is greater with **a slow beam.**

Vertical and Horizontal Plates

If two sets of deflection plates (Fig. 4-27) are placed at right angles to each other inside a crt, the electron beam can be controlled in any direction.

By varying the potential of the vertical-deflection plates, the spot on the face of the tube can be made to move up and down. The distance will be proportional to the change in potential between the plates. Changing the potential difference between the horizontal-deflection plates will cause the beam to move a given distance from one side to the other. There are also other directions than up–down and left–right as the beam must be deflected in all directions.

Study the two diagrams in Fig. 4-28. You should be able to see that the beam can be moved to any position on the screen, simply by moving it both vertically and horizontally.

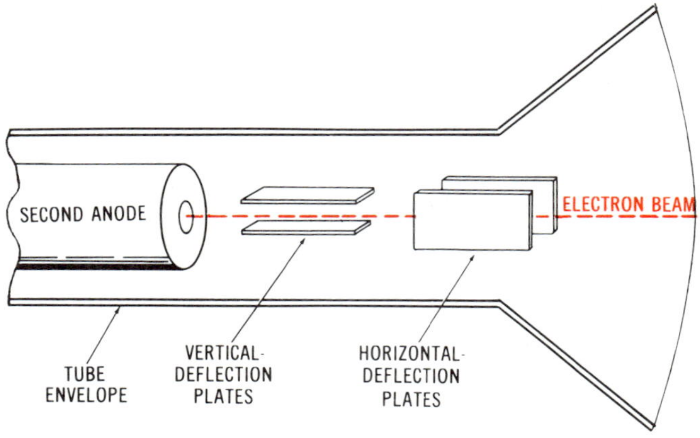

Fig. 4-27. Deflection-plate arrangement.

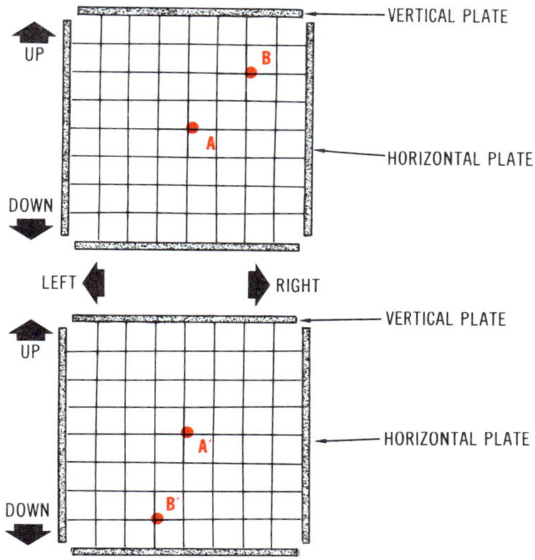

Fig. 4-28. Diagrams illustrating electron beam movement.

In the top diagram (Fig. 4-28), position A of the beam is in the center. It can be moved to position B by going up two units and then right two units. Movement of the beam is the result of the simultaneous action of both sets of deflection plates. The electrostatic field between the vertical plates moves the electrons up an amount proportional to two units at the screen. As the beam passes between the horizontal plates, it is moved to the right an amount proportional to two units at the screen.

If the amount of deflection in Problem Q4-29 occurred to the left and down so that each set of plates acted at the same time, the picture would be like the one shown in Fig. 4-29A. For example, if the vertical plates moved the beam downward

Q4-27. Which of the methods in Q4-22 to Q4-26 would be used to change a crt's deflection angle?

Q4-28. In the bottom diagram in Fig. 4-28, how many units and in which direction will each set of deflection plates move the beam from A′ to B′?

Q4-29. Roughly draw a line on a graph to represent the picture seen on the screen as the spot moves from A′ to B′ (in the bottom diagram) in Fig. 4-28.

Your Answers Should Be:

A4-27. **Varying the potential on the deflecting plates** would be the method used to change the deflection angle in a crt.

A4-28. The vertical plates will move the beam **down three units.** The horizontal plates will move the spot **one unit to the left.**

A4-29. The picture on the screen will look like either:

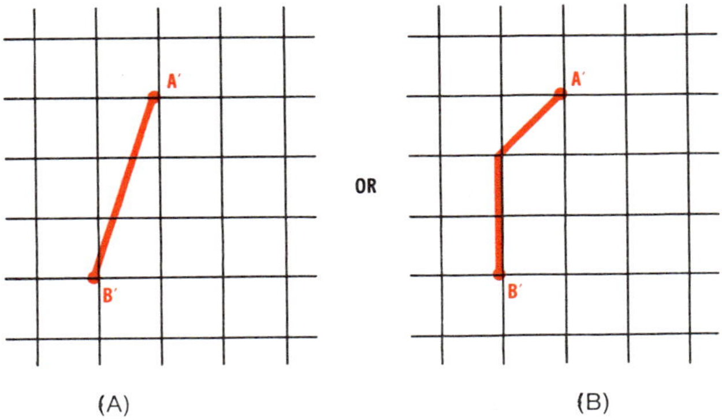

Fig. 4-29.

at the rate of 1 unit per second and the horizontal plates moved it to the left at the rate of ⅓ unit per second, both movements would have been completed in 3 seconds at point B′. The result would be a straight line.

In the example on the right, the potential on the vertical and horizontal plates changes at the same rate. In the same time period—say 1 second—both plates move the beam 1 unit. The horizontal plates have completed their task at the end of 1 second, but the vertical plates have moved the beam only ⅓ of the required distance. If this were true, the picture shown in Fig. 4-29B would appear on the screen.

Amplitude Versus Time

Do you recall the statement made earlier that waveforms could be described in terms of amplitude and time? You have just seen how the movement of the crt beam depends on both potential (amplitude) and time.

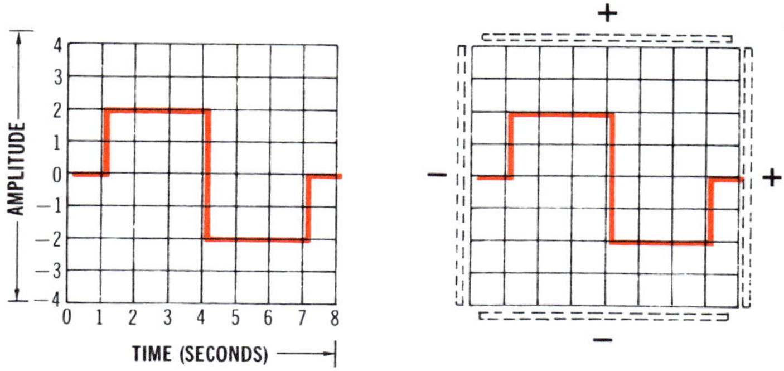

(A) Amplitude-time graph. (B) Crt-screen presentation.

Fig. 4-30. Amplitude versus time.

From zero time to 1 second, the waveform in the diagrams of Fig. 4-30 is at zero volts. In the crt presentation, the vertical plates remain at the same potential difference while the potential difference between the horizontal plates increases 1 unit in the direction necessary to move the beam toward the right. When time is equal to 1 second, the waveform rises to +2 volts. The potential difference between the vertical plates increases enough to move the electron beam 2 units in the positive direction. From 1 to 4 seconds, the waveform remains at +2 volts and then decreases to −2 volts. As the horizontal-plate potential difference increases by 3 units, the vertical potential remains the same (+2 units) and then drops sharply 4 units (Fig. 4-30B). For the next 3 seconds, the waveform remains at −2 volts. In the crt, the potential difference between the vertical plates remains unchanged as the horizontal potential increases uniformly by 3 units.

The vertical-plate potential difference follows the **voltage** of the waveform. The horizontal-plate potential follows the passage of **time.** Together they determine the **trace** (image produced on the screen by the moving beam).

Q4-30. Waveforms can be described in terms of ____ and __________.

Q4-31. The horizontal-deflection plates are used to reproduce the ____ __________.

Q4-32. The vertical-deflection plates are used to reproduce the __________ __________.

Your Answers Should Be:

A4-30. Waveforms can be described in terms of **time** and **amplitude.**

A4-31. The horizontal-deflection plates are used to reproduce the **time component.**

A4-32. The vertical-deflection plates are used to reproduce the **amplitude component.**

Voltage Control of Horizontal Plates

Assume that the resistance of the potentiometer, shown in Fig. 4-31, is spread evenly along its length. When the arm of the potentiometer is at the middle position, there is the same potential on each plate. Since there is zero potential difference

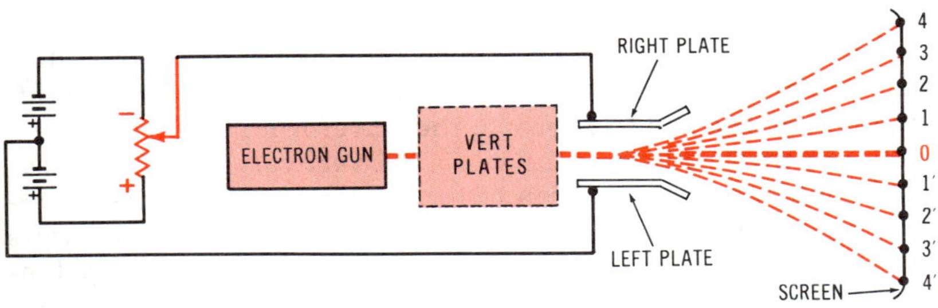

Fig. 4-31. Horizontal plates—top view.

between the plates, an electrostatic field is not produced. The beam will be at zero on the screen. If the arm is moved downward at a uniform rate, the right plate will become more positive than the left. The electron beam will move from 0 through 1, 2, 3, and 4 in equal time intervals. Then, if the potentiometer arm is moved at the same rate in the opposite direction, the right plate will decrease in positive potential. The beam returns to the zero position when the potential difference between the plates again becomes zero. Moving the arm of the potentiometer toward the other end of the resistance will cause the left plate to become more positive than the right. The direction of the electric force reverses, and the beam moves from 0 through 4′. If the movement of the potentiometer arm is at a linear (uniform) rate, the beam will move at a steady rate.

The ends of the deflection plates are bent outward to permit wide-angle deflection of the beam. The vertical plates are bent in the same manner.

Moving a potentiometer arm is satisfactory for purposes of illustration but, in real oscilloscopes, this is not a practical way to vary the horizontal-deflection voltage. Nearly all oscilloscopes with electrostatic deflection use a sawtooth waveform applied to the horizontal plates to produce horizontal deflection of the beam.

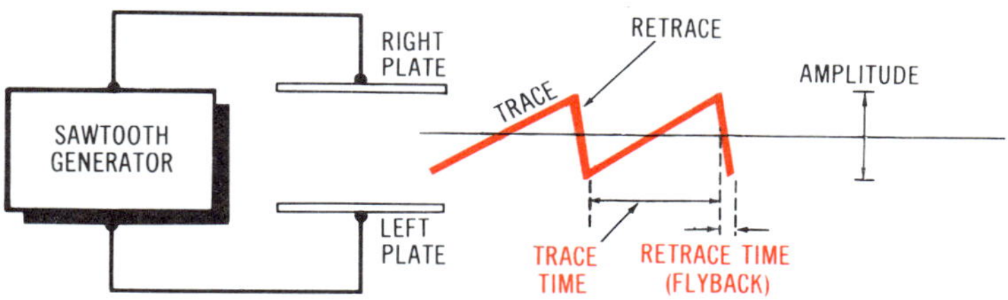

Fig. 4-32. Using a sawtooth waveform to produce horizontal beam deflection.

At the reference line, the potential on both plates is equal. Below the line, the waveform makes the left plate more positive and, above the line, the right plate is made more positive than the other. The waveform amplitude causes a uniform movement of the beam across the screen. The retrace line (trailing edge of the waveform) brings the beam quickly back to the starting point.

Q4-33. How do most oscilloscopes obtain a linear rate of deflection for use as a time base?

Q4-34. Why are the ends of the deflection plates bent outward?

Q4-35. In the illustration of Fig. 4-31, what is the potential difference between the deflection plates when the potentiometer is centered?

Q4-36. What is the positive-going section of the sawtooth waveform called?

Q4-37. What is the negative-going section of the sawtooth waveform called?

Your Answers Should Be:

A4-33. Most oscilloscopes use a **sawtooth waveform** applied to the horizontal plates of the crt.

A4-34. They are bent outward **to permit wide-angle deflection of the beam.**

A4-35. The potential difference is **zero.**

A4-36. It is called the **trace portion.**

A4-37. It is called the **retrace portion.**

Crt Graticule

It is possible to cover the face of the crt with a sheet of plastic on which are scribed horizontal and vertical lines. This marked plastic sheet is called a **graticule.**

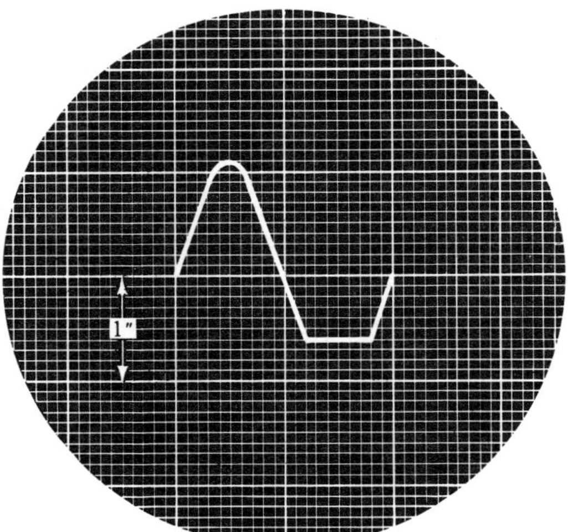

Fig. 4-33. Screen of cathode-ray tube.

The graticule can be used to determine the voltage of waveforms because the **deflection sensitivity** of a crt is uniform throughout the vertical plane of the screen. Deflection sensitivity is a constant which is dependent on the construction of the tube. It states the number of inches, centimeters, or millimeters that the beam will be deflected for each volt of potential difference applied to the deflection plates. Deflection sensitivity is directly proportional to the physical length of the deflection plates and their distance from the screen. It is inversely proportional to the distance between the plates and to the second-anode voltage.

Deflection sensitivity for a given crt might be 0.2 millimeter (mm) per volt. This means the spot on the screen will be deflected 0.2 mm (about 0.008 inch) when a difference of 1 volt exists between the plates. Sometimes the reciprocal of deflection sensitivity (called **deflection factor**) is given. The deflection factor for the example given would be 1/0.008, or 125 volts per inch. Sensitivity is usually measured in inches per volt and deflection factor in volts per inch.

In the preceding example, 125 volts applied between one set of plates would deflect the beam 1 inch on the screen. This means that the deflection caused by small signals could not be observed. For this reason, the deflection plates are connected to amplifiers that magnify the signals.

Assume that a peak-to-peak value of a known voltage applied to the oscilloscope indicates that each inch marking on the graticule is equal to 60 volts. Each of the ten subdivisions will therefore have a value of 6 volts. Most oscilloscopes have controls to **attenuate** (decrease) or increase the strength of a signal before the signal is placed on the deflection plates. **Attenuator** and **gain-control** settings must not be disturbed after the calibration has been made. For maximum accuracy, recalibrate the graticule each time a voltage is to be measured.

Q4-38. If the graticule in Fig. 4-33 has been calibrated to 50 volts per inch, what are the values of the positive and negative peaks of the waveform?

Q4-39. What is deflection sensitivity?

Q4-40. In what units is deflection sensitivity measured?

Q4-41. What is the reciprocal of deflection sensitivity called?

Q4-42. Deflection sensitivity is inversely proportional to what? It is directly proportional to what?

Q4-43. How is a signal magnified for screen presentation?

Q4-44. What must be done for maximum accuracy each time a voltage is to be measured with a graticule?

Your Answers Should Be:

A4-38. The values are **55 volts** for the positive peak and **30 volts** for the negative peak.

A4-39. Deflection sensitivity is **the distance that the spot on the screen will be deflected for each volt of potential difference applied to the deflection plates.**

A4-40. Deflection sensitivity is measured in **inches per volt.**

A4-41. **Deflection factor** is the reciprocal of deflection sensitivity.

A4-42. Deflection sensitivity is inversely proportional to the **distance between the plates and the accelerating voltage on the second anode.** It is directly proportional to the **length of the deflection plates and the distance from the plates to the screen.**

A4-43. **Amplifiers** magnify the signal.

A4-44. **The graticule must be recalibrated.**

Crt Designation

Cathode-ray tubes are designated by a tube number, such as 2AP1, 2BP4, 5AP1A, etc. The first number identifies the diameter of the tube face. Typical diameters are 2 inches, 5 inches, and 7 inches. Tubes can have diameters up to 24 inches or more. The first letter designates the order in which a tube of a given diameter was registered. The letter-digit combination indicates the type of phosphor (glowing material) used on the screen. Phosphor P1, which is used in most oscilloscopes, produces a green light at medium **persistence.** P4 phosphor provides a white light and has a short persistence. Persistence refers to the length of time the phosphor glows after the electron beam is removed. If a letter appears at the end, it signifies the number of the modification after the original design.

Crt Safety

Handle the cathode-ray tube with a great deal of care. Because of its size and air-evacuated condition, a tremendous amount of pressure is exerted inward over all its surface. A bump or even a scratch may weaken the glass, causing it to **implode** (opposite of explode but with the same results). Pieces of glass and parts will fly in all directions. When replacing a crt, it is a good practice to wear safety glasses and gloves. Also, you should store the old tube in the box in which the new tube came in for safely disposing of it later.

CONTROL CIRCUITS

Although the cathode-ray tube is a highly versatile device, it cannot operate without control circuits. Naturally, the type of control circuits required depends on the purpose of the equipment in which the crt is used.

There are many different types of oscilloscopes. They vary in purpose and cost—from relatively simple test instruments to highly accurate laboratory models. However, all have two things in common; they must have some type of crt, and they must have a group of control circuits to permit feeding a waveform to the crt. Although there are other types of circuits, most test oscilloscopes can be divided into the basic sections as shown in Fig. 4-34.

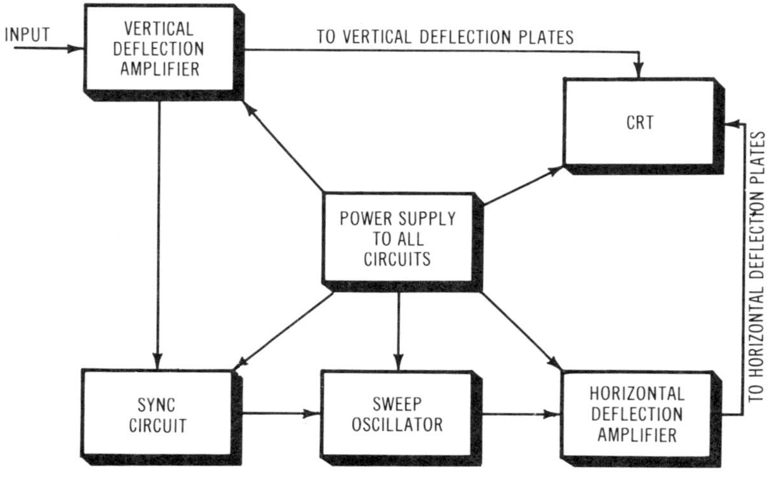

Fig. 4-34. Basic sections of an oscilloscope.

Q4-45. What does a 3AP1B crt number designate?

Q4-46. What must all oscilloscopes have in common?

Your Answers Should Be:

A4-45. The number 3AP1B designates a crt **3 inches in diameter** and the **first of its diameter registered.** Its trace is **green with medium persistence,** and the crt is the **second modification** of the original.

A4-46. All oscilloscopes have in common **a crt and a group of control circuits.**

Front-Panel Controls

There are several front-panel controls used to adjust the oscilloscope circuits for proper operation. The type and number of controls vary with the purpose of the **scope** (an accepted name for oscilloscope). The following pages will discuss these controls, in conjunction with the circuits identified in Fig. 4-34. Some typical controls are shown in Fig. 4-35.

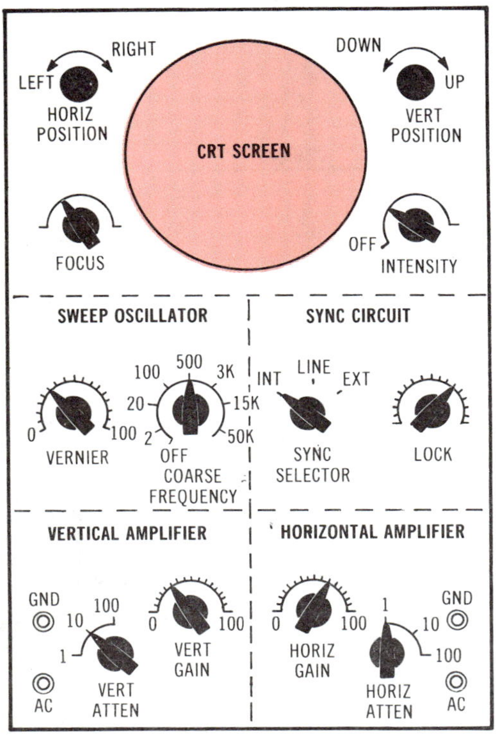

Fig. 4-35. Oscilloscope panel.

All of the circuits in the block diagram of Fig. 4-34 are represented on the front panel. (The power-supply switch is on the intensity control.) The four controls surrounding the screen regulate voltages being fed to the crt. The four areas in the lower half of the panel carry titles similar to those used in the block diagram.

Power Supply

Power-supply requirements for oscilloscopes vary considerably. Certain cathode-ray tubes require accelerating (second anode) voltages as high as 15 to 30 KV (15,000 to 30,000 volts). The type used with the general-purpose scope, on the other hand, uses 1 to 3 KV. Most power supplies employ a transformer, half- or full-wave rectifiers, filters, a load resistance, and, in some cases, voltage regulation.

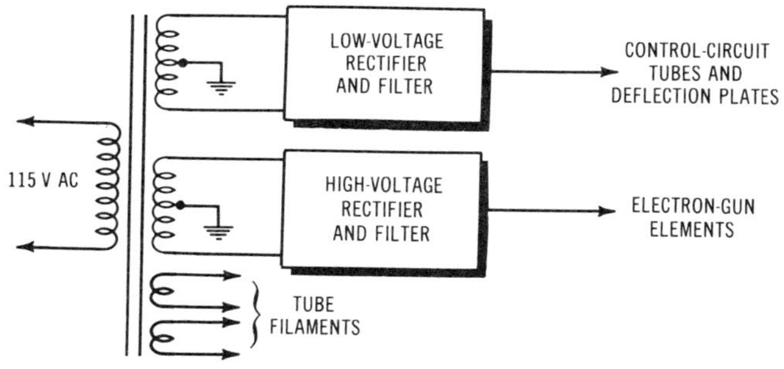

Fig. 4-36. Typical power supply.

Most test scopes have both high-voltage and low-voltage power-supply sections fed by a single transformer. The high-voltage, low-current section takes care of the electron-gun requirements. Voltage needs for the remainder of the circuits are supplied by the low-voltage section. This section may provide potentials as high as 300 or 400 volts. A third or fourth winding on the transformer provides voltage and current for the vacuum-tube heaters.

Q4-47. To which element of the cathode-ray tube is the INTENSITY control connected?

Q4-48. To which element of the crt is the FOCUS control connected?

> **Your Answers Should Be:**
>
> **A4-47.** The INTENSITY control is connected to the **control grid** of the crt.
>
> **A4-48.** The FOCUS control is connected to the **first anode.**

Crt Controls

In the circuit of Fig. 4-37, the second anode (accelerator) is at ground potential. To obtain the high accelerating potential

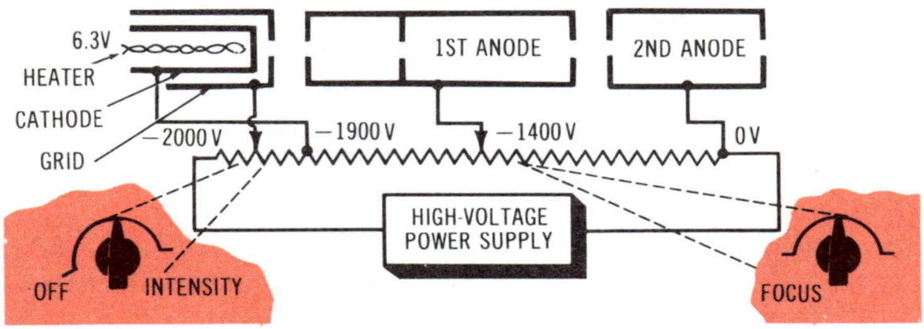

Fig. 4-37. Electron gun.

required, the other electron-gun elements are operated at negative potentials. The control grid normally operates near 2000 V negative, or 90 to 100 V more negative than the cathode. The first anode (focusing) can be maintained between −1200 and −1600 V. These voltages are typical but vary among different instruments.

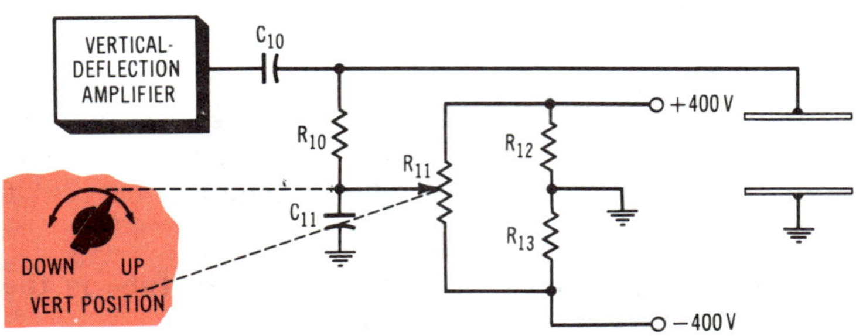

Fig. 4-38. Vertical positioning.

Deflection-Plate Controls

The following method (Fig. 4-38) of adjusting the deflection-plate voltage is only one of several possible ways.

In addition to centering the beam vertically on the screen, there are times when it is desirable to move the entire waveform up or down. VERT POS (vertical positioning) is a front-panel control that permits this. A circuit used to vary the potentials on the plates, for positioning purposes, is shown in Fig. 4-38. Voltage from the last stage of the amplifier, varying in the same way as the original waveform, is impressed across R_{10}. Capacitor C_{11} returns the ac signal to ground and blocks dc.

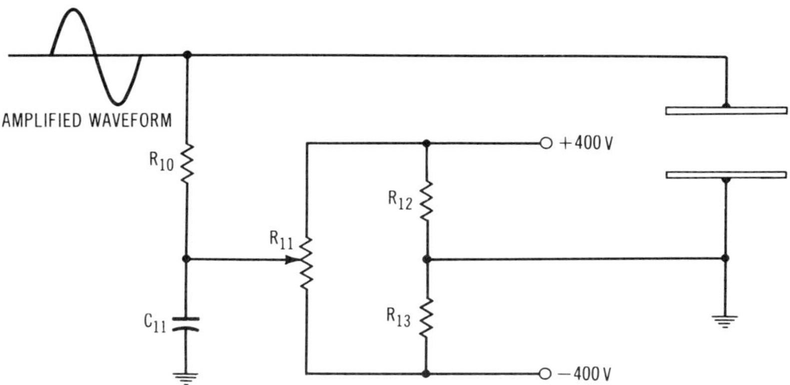

Fig. 4-39. Positioning circuit.

When R_{11} (VERT POS control) is centered, there is no difference of potential between the two plates. When the arm is moved down, the lower plate becomes more positive than the upper plate, and the electron beam moves downward. When the arm is moved up, the upper plate becomes more positive. If there is a waveform being applied across R_{10} (Fig. 4-39), the difference of potential from this positioning network is added to or subtracted from it. This arrangement makes it possible to shift the entire waveform up or down on the crt screen.

Q4-49. If one deflection plate is at +124 V and the other is +18 V, in which direction will the electron beam bend?

Your Answers Should Be:

A4-49. The electron beam will bend **toward the +124 V plate.**

Horizontal Positioning

One type of horizontal-positioning circuit used in a deflection system is shown in Fig. 4-40. The positioning transistor,

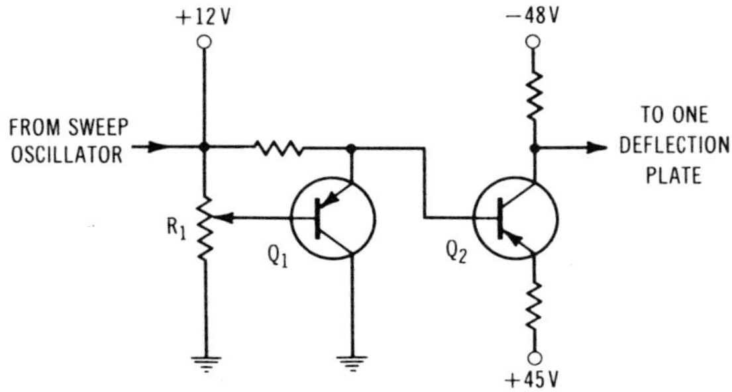

Fig. 4-40. Horizontal-positioning circuit.

Q_1, operates as an emitter follower. Its input is a sawtooth sweep voltage from the sweep-oscillator circuit. Variable resistor R controls the base voltage of the emitter-follower transistor Q_1, which is dc-coupled from its emitter to the base of deflection amplifier Q_2. The dc component of the deflection voltage at the output from the collector of transistor Q_2 is applied to one of the crt deflection plates. Another amplifier stage similar to that of transistor Q_2 drives the other deflection plate. The difference in dc voltage between the two deflection plates, which is set by variable resistor R, determines the position of the trace on the oscilloscope screen. Vertical positioning is done similarly.

Vertical Amplifier

Since the vertical amplifier receives the waveform to be observed, its input impedance should be very high to prevent loading of the external circuit from which the waveform is obtained and to prevent the resultant distortion of the signal. The amplifiers of most scopes have input impedances of several megohms. Some other requirements for good vertical amplifiers are listed below.

Frequency Response—Frequency response is a measure of the ability of an amplifier or circuit to pass all of the frequency components of a waveform. A pure sine wave (Fig. 4-41A), as you know, has only one frequency component—the **fundamental frequency.**

(A) Sine wave. (B) Square wave.

Fig. 4-41. Frequency components of a waveform.

A square wave, however, consists of the fundamental sine wave plus many odd-numbered **harmonics** (Fig. 4-41B). A harmonic is a sine wave having a frequency that is a whole-number multiple of the fundamental frequency. A perfect square wave has an infinite number of odd-numbered harmonics. Its tops and bottoms are perfectly flat, and the rise and decay of its sides occur in zero time. Since there must be some time to allow voltages to rise and fall, there is no practical circuit that can produce a perfect square wave. However, a conventional square wave contains several hundred odd-numbered harmonics.

A good general-purpose scope should have a frequency response extending up to 2 megahertz. For practical maintenance work, a scope should be able to pass the tenth odd harmonic of a square wave. Since this is 21 times the fundamental frequency, a 2-megahertz scope should be able to display square waves having a fundamental frequency as high as 100 kHz.

Each of the pulse waveforms shown in Fig. 4-42 consists of a different combination of fundamental and harmonic frequen-

Q4-50. Why are the deflection plates of a crt bent outward at the end?

Q4-51. What makes a square wave different from a sine wave?

Your Answers Should Be:

A4-50. The ends of the plates are bent to **allow larger angles of electron-beam deflection** than would be permitted by straight plates.

A4-51. A sine wave is made up of a single fundamental frequency; a square wave consists of the **fundamental frequency plus many odd harmonics** of the fundamental.

cies. In order to display such waveforms accurately, a scope must have a good high-frequency response. This is so that the higher harmonics will be amplified the same amount as the fundamental and the lower harmonics.

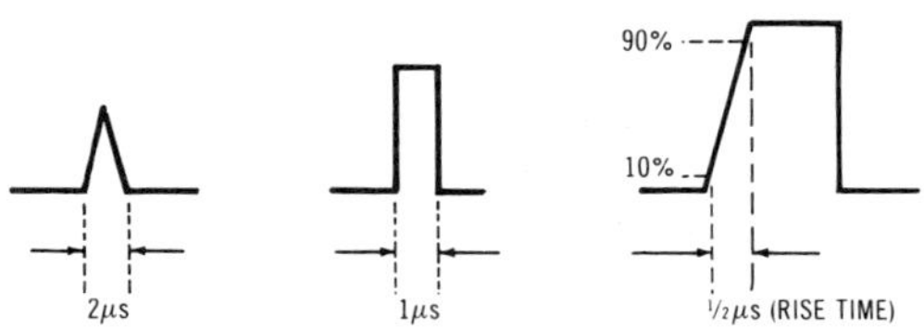

Fig. 4-42. Pulse duration.

Another way to examine the response of a scope is in terms of **rise time.** The rise time is the time between the 10% and 90% amplitude points on the leading edge of a pulse. The minimum rise time that a scope can reproduce is determined mainly by the charge time of certain capacitances in the scope.

Gain—The gain of a vertical amplifier determines how well a small signal can be expanded for observation on the screen. If the crt, for example, has a deflection factor of 0.8 V per inch and no means of amplification, a waveform having 0.2 V amplitude would be very difficult to examine. However, if an amplifier were used, all large signals would be amplified so much that they would extend off the screen. Therefore, instead of having several channels of amplification (each with its set of linear, good frequency-response amplifiers), a method must be used to attenuate (reduce) waveform amplitudes before they arrive at a single channel of amplification.

The diagram given in Fig. 4-43 shows one method often used

for attenuation. The **VERT ATTEN** (vertical attenuator) switch has three positions, 1, 10, and 100, which are factors of attenuation. The attenuation equals unity in position 1; there is no attenuation of signal. This corresponds with the top tap of the switch shown in the schematic. The full voltage of the input is fed to the gate of the source-follower. Attenuation equals 1/10 in position 10. Resistors R_1, R_2, and R_3 are selected so that 1/10 of the input voltage will arrive at the gate. Position 100 provides an attenuated signal of 1/100.

Since attenuation values between these broad settings may be desired, a finer attenuation control is provided. This is the **VERT GAIN** (vertical gain) control. As you can see, it selects a voltage from R_4, which is part of the source resistance, and applies this voltage to the vertical amplifier. Through the use of the VERT ATTEN and VERT GAIN controls, the vertical size of the waveform can be regulated on the screen.

The vertical-deflection amplifier stage in a good scope is usually a push-pull amplifier having a constant gain and a frequency response up to 2 MHz. The output of the amplifier is fed to the vertical-deflection plates.

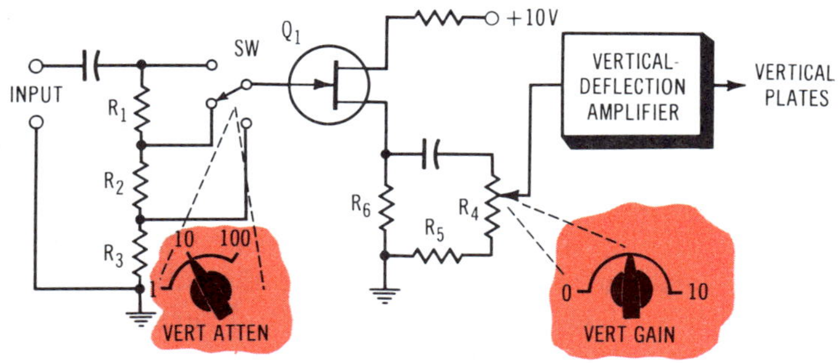

Fig. 4-43. Attenuation circuit.

Q4-52. Constant gain refers to the ability of an amplifier to equally amplify all signals within its capability. Why is this necessary in an oscilloscope?

Q4-53. Is the frequency response of a scope a good measure of its capabilities?

Q4-54. In the schematic of Fig. 4-43, would the switch be connected to the tap at the bottom of R_2 or the top of R_1 if the VERT ATTEN were set at 100?

Your Answers Should Be:

A4-52. Constant gain is required so that **all waveforms,** regardless of their amplitude (within the voltage range of the amplifier), **are amplified the same amount.** Variations in gain would make the scope presentations inaccurate.

A4-53. **Yes,** frequency response is a good measure of the capability of a scope. A scope with a good frequency response will reproduce waveforms over a wider frequency range more faithfully (with less distortion) than a scope with a poorer frequency response. A scope with a good frequency response responds more quickly to the rapid changes of narrow pulses and steep wave slopes.

A4-54. The switch would be connected to the **bottom tap,** thus providing the grid with less signal voltage than at the other two taps.

Other Vertical-Amplifier Requirements

Inputs to the Scope—The illustration of the front panel of an oscilloscope shows GND and AC connections for the vertical-deflection amplifier. Test leads with probes attached are inserted into these connections for test purposes. On some oscilloscopes, there is a third jack that is marked DC. This provides the possibility of observing a dc voltage or a waveform that varies its amplitude at a very slow rate. The DC connection feeds the signal directly to a dc amplifier and, then, to the deflection plates. The normal vertical amplifiers cause distortion of very low-frequency signals.

Y-Axis Amplifier—On some scopes, the vertical-deflection amplifier is called a **Y-axis amplifier.** The Y axis corresponds to the Y coordinate (up-and-down reference line) on a graph. Since a scope presents a graph of amplitude (plotted on the Y axis) and time (plotted on the X axis), these terms are sometimes used instead of vertical and horizontal.

If the vertical amplifier and its associated circuits are properly designed according to the requirements you have just studied, the amplitude of a waveform will be faithfully reproduced on the screen.

An amplifier is required to increase signal voltages so that as much of the full size of the screen as possible can be used. It is easier and more accurate to study an enlarged reproduction of a waveform. Large waveforms can be attenuated to

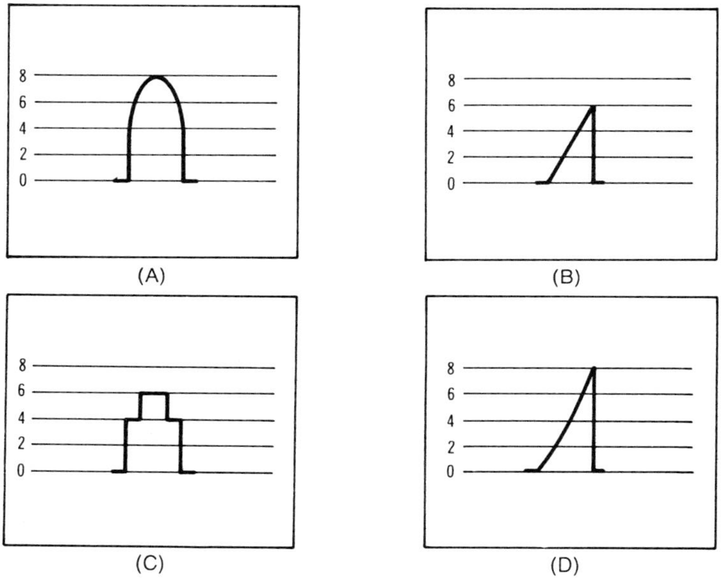

Fig. 4-44. Sketches for Problem Q4-55.

1/10 or 1/100 of the amplified size, and any waveform can be made larger or smaller by varying the amplifier gain.

Q4-55. **Assume your oscilloscope had only a crt, vertical-deflection amplifier, and the right type of power supply. Draw a picture of each of the above waveforms, showing how they might appear on the screen.**

Q4-56. **What are the two characteristics of a waveform that an oscilloscope is able to reproduce?**

Q4-57. **____ is plotted on the X axis, and __________ is plotted on the Y axis.**

Q4-58. **Why is a DC input jack or terminal included on some oscilloscopes?**

Q4-59. **What other vertical-amplifier inputs are used in an oscilloscope?**

Your Answers Should Be:

A4-55. Your drawings should look something like the sketches in Fig. 4-45.

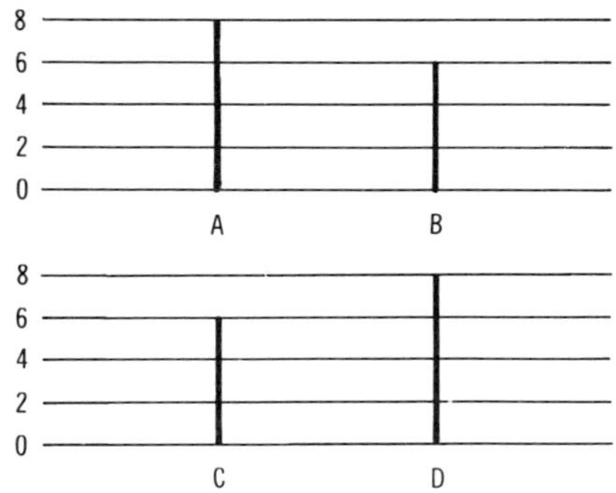

Fig. 4-45. Sketches for Answer A4-55.

A4-56. The two waveform characteristics that an oscilloscope can reproduce are **amplitude** and **time.**

A4-57. **Time** is plotted on the X axis, and **amplitude** is plotted on the Y axis.

A4-58. The DC input jack is used to **observe a dc voltage or a waveform that varies in amplitude at a low rate.**

A4-59. **AC** and **GND** (ground) connections are the other two inputs.

Horizontal Time Base

As you can see in Answer A4-55, a scope with only a vertical-deflection amplifier in its control circuits will present only a vertical line; the horizontal dimension is missing.

Time as a Reference—Since waveforms change their amplitude in accordance with time, it becomes a useful means of measurement for the horizontal direction on the screen.

Look at Fig. 4-46. If the two waveforms span the same period of time, each could be divided into corresponding increments (small intervals) of time. If a sawtooth waveform was applied to the horizontal plates and a sine wave was applied to the vertical plates, the former would move the electron beam sideways, and the latter would move it up or down in corresponding increments of time. Notice how the vertical and hori-

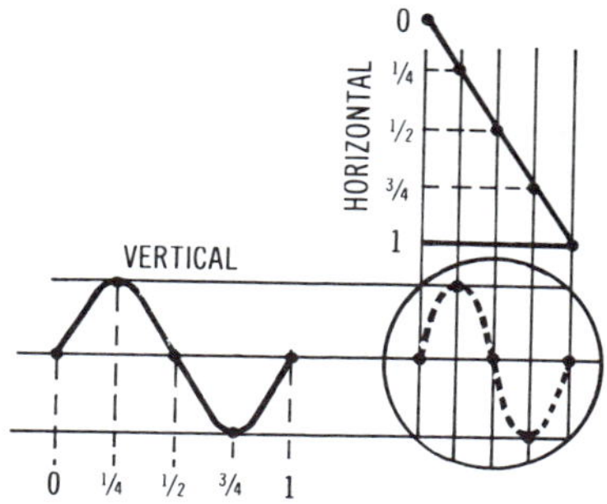

Fig. 4-46. Rate of change.

zontal deflections combine at each instant of time to produce the waveform.

Characteristics of a Sawtooth Waveform—You have probably identified the necessary characteristics of a sawtooth waveform. Voltage must rise uniformly to be constantly proportional to time. It must be capable of starting its rise at the same instant that the waveform being observed starts. The time duration of the sawtooth waveform must be equal to that of the other waveform if one complete cycle is to be observed. The sawtooth must decay quickly to zero so that both waveforms can complete their cycles at the same time.

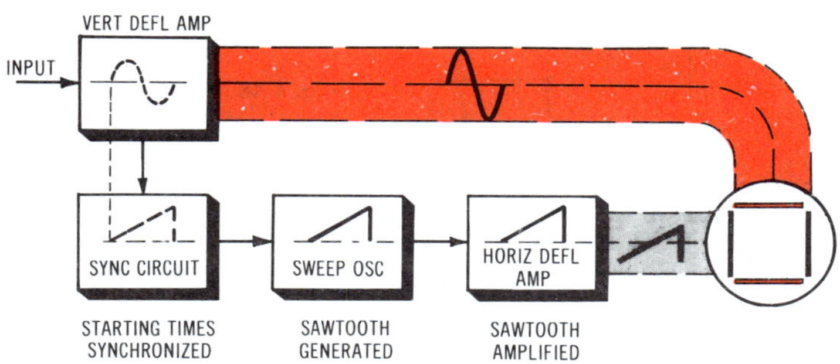

Fig. 4-47. Vertical and horizontal deflection.

Q4-60. The _ _ _ _ _ _ _ _ _ waveform moves the electron beam from side to side, and the _ _ _ _ _ _ _ _ _ _ _ waveform moves the beam up and down.

Q4-61. What part of the control circuits of a scope produces all the characteristics of the sawtooth waveform?

Your Answers Should Be:

A4-60. The **sawtooth** waveform moves the electron beam from side to side, and the **sinusoidal** waveform moves the beam up and down.

A4-61. **The horizontal-deflection circuits** produce all the characteristics of the sawtooth waveform.

Sweep-Oscillator Circuits

The sawtooth waveform is generated by the sweep oscillator. Sweep refers to the steady rise of sawtooth voltage that moves the waveform horizontally across the screen in a desired period of time. An oscillator is a circuit that is capable of repeating the waveform which it generates at some specific frequency.

In ac fundamentals, you learned about the simple RC circuit shown in Fig. 4-48. The circuit contains a resistor and a capacitor in series with a battery. A switch capable of disconnecting the battery and placing a short circuit across R_1 and C_1 is also connected in this circuit.

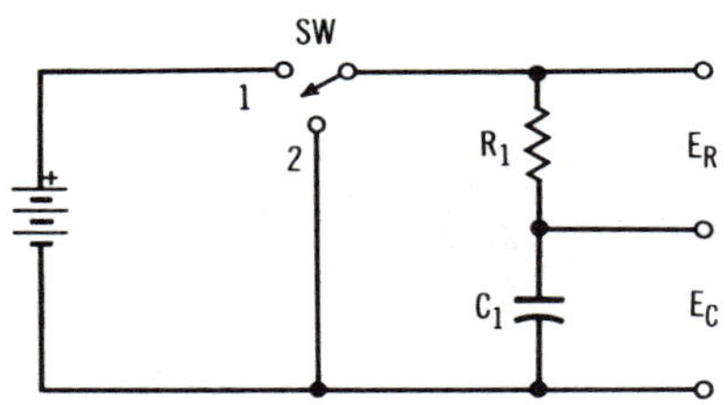

Fig. 4-48. Simple RC circuit.

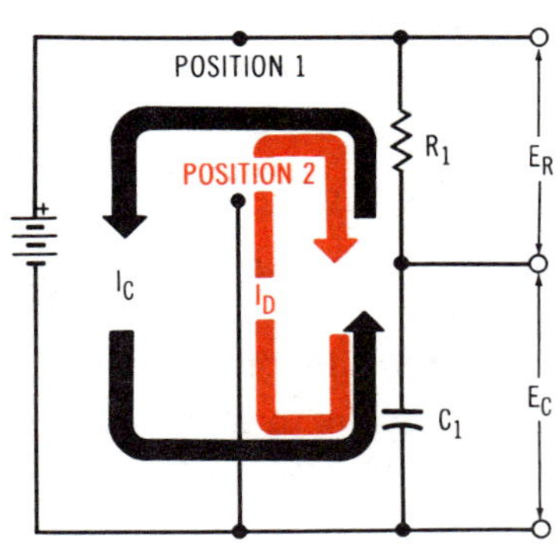

Fig. 4-49. Current flow in the circuit of Fig. 4-48.

At the instant that the switch is placed in position 1 (Fig. 4-49), I_C (charge current) rises to maximum, and E_R rises to the value of the battery voltage (Fig. 4-50). As capacitor C_1 charges (E_C) at an exponential rate, I_C and E_R decrease at the same rate. At the end of a period of time determined by the values of R_1 and C_1, the capacitor will reach its maximum charge. Current will stop flowing and E_R will become 0. At time 2 (Fig. 4-50), when the switch is in position 2, the capacitor begins to discharge. I_D (discharge current) is maximum negative (reverses direction), and E_R is also maximum in the negative direction (Fig. 4-49). The discharge decreases exponentially until all values reach 0. E_C resembles the sawtooth, but its rise is not linear.

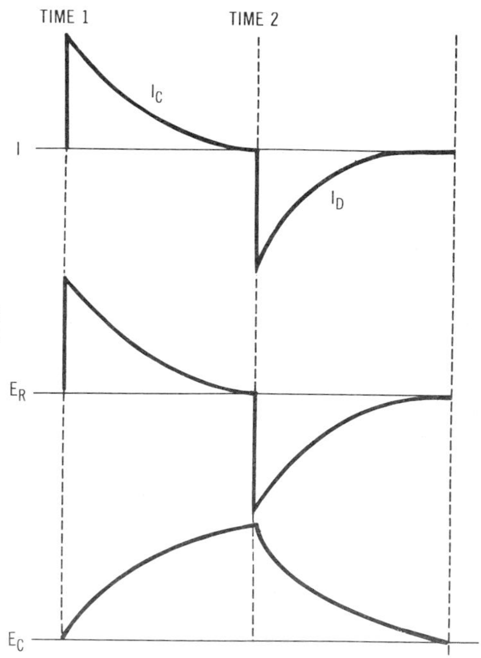

Fig. 4-50. Waveforms showing voltage and current action in Fig. 4-48.

Q4-62. Refer to Fig. 4-48. At the instant that the switch is placed in position 1, is E_R equal to, greater than, or less than E_C?

Q4-63. Assume C_1 has been charged. At the instant the switch is placed in position 2, is E_R equal to, greater than, or less than E_C?

Your Answers Should Be:

A4-62. **E_R is greater than E_C** at the instant the switch is closed. E_R is at maximum voltage, and E_C is at zero.

A4-63. **E_R is less than E_C** at the instant the switch is placed in position 2. When C_1 is fully charged, there will be zero volts across the resistor, and the voltage across C_1 will be at its maximum.

Developing a Sawtooth Waveform

There are several types of sawtooth generating circuits—neon tube, thyratron, multivibrator, etc. The **thyratron sawtooth generator** is representative of how all of these circuits operate and it is easy to understand.

A thyratron is a triode containing an ionizing gas. In Fig. 4-51, B+ current flows through R_1 and charges C_1. When the voltage across C_1 reaches a certain potential, the gas ionizes.

When this happens, the thyratron conducts and rapidly discharges the capacitor. The voltage across C_1 has a waveform that depends on the charge and discharge times. Charge time can be lengthened by increasing the value of R_1, C_1, or both. The bias on the grid also controls the time at which the tube conducts. A larger negative grid-to-cathode potential, making it more difficult for the tube to conduct, will require a larger ionizing potential on the plate. It will take the RC circuit longer to reach this potential, and the charge time of the sawtooth waveform will be longer. When the tube conducts, the capacitor discharges until a voltage across the tube is reached that no longer supports ionization. Capacitor C_1 then recharges and the cycle repeats.

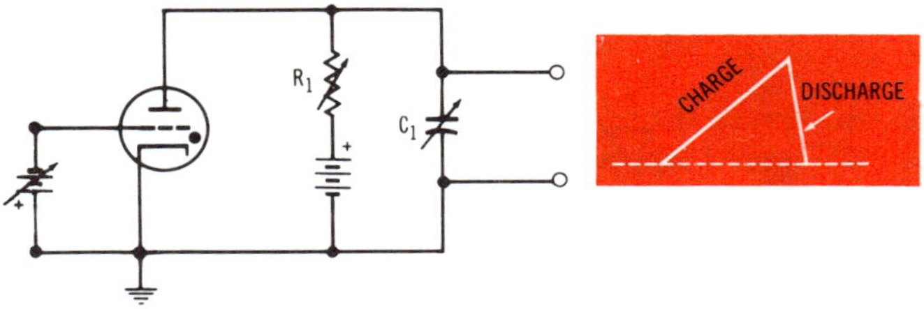

Fig. 4-51. A thyratron circuit.

As can be seen in Fig. 4-52, the capacitor will charge until it accumulates a voltage equal to the ionizing potential of the thyratron. The tube conducts current and discharges the capacitor. Since the thyratron acts as a short across the capaci-

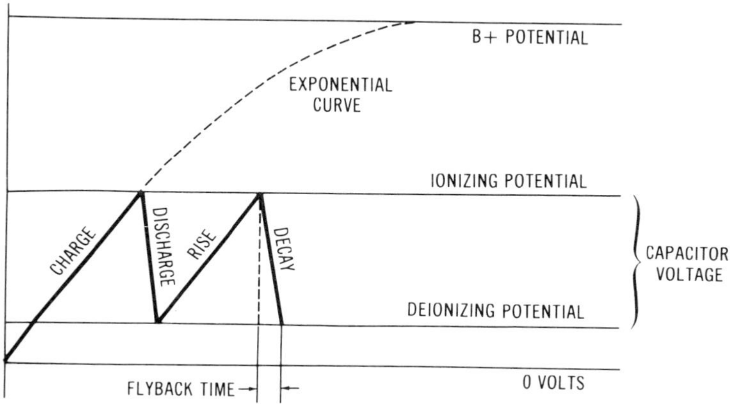

Fig. 4-52. Thyratron linearity.

tor, the capacitor discharges rapidly. When the capacitor voltage falls to the **deionizing potential** of the tube, the thyratron stops conducting and acts as an open switch. Then the cycle repeats. The charge of the capacitor corresponds to the rise of the sawtooth waveform, and the discharge corresponds to its decay.

The linearity of the sawtooth voltage across the capacitor is determined by the ionizing and deionizing action of the thyratron. The tube discharges the capacitor while its charge voltage is in the lower, more linear part of the exponential curve. The thyratron is also capable of discharging the capacitor rapidly, keeping decay time of the waveform to a minimum. If decay, or **flyback time** as it is most often called, is too long, horizontal deflection will not return to the starting point before the waveform on the vertical plates has started its next cycle.

Q4-64. What are some types of sawtooth-generating circuits used in electronics?

Q4-65. What is a thyratron?

Q4-66. How can the charge time of the thyratron circuit be controlled?

Your Answers Should Be:

A4-64. **Neon-tube, thyratron,** and **multivibrator** circuits.

A4-65. A thyratron is a **triode containing an ionizing gas.**

A4-66. It can be controlled by varying the **grid bias of the tube** or the **values of R_1 and C_1.**

A Typical Sawtooth Generator

Since the frequencies, or time durations, of waveforms are not all the same, a sawtooth waveform with only a single rise time is not suitable. The most used method of varying the length of a sawtooth waveform is to change the values of the RC charging circuit.

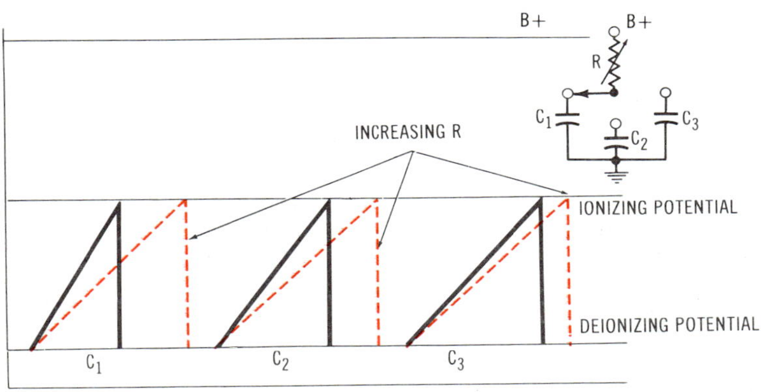

Fig. 4-53. Changing length of sawtooth waveform.

By changing the capacitors in the RC circuit, the RC time constant can be increased in coarse increments, as shown by the solid lines in Fig. 4-53. Capacitor C_1 has a smaller capacitance than C_2, which is smaller than capacitor C_3. If resistance R remains the same, a larger capacitance will take longer to charge than a smaller one. Consequently, the rise time of the sawtooth waveform generated by the capacitor would increase. If a variable resistor is used for R, fine variations of the basic sawtooth waveform, for each value of C, can be made and controlled. This is shown by the dashed lines in Fig. 4-53. In each case, the firing potential (which determines the time at which the capacitor discharges) would remain the same.

Fig. 4-54 shows one version of the thyratron sawtooth generator that is used in oscilloscopes. The cathode is maintained at a small positive voltage (about 3 V) by the voltage divider

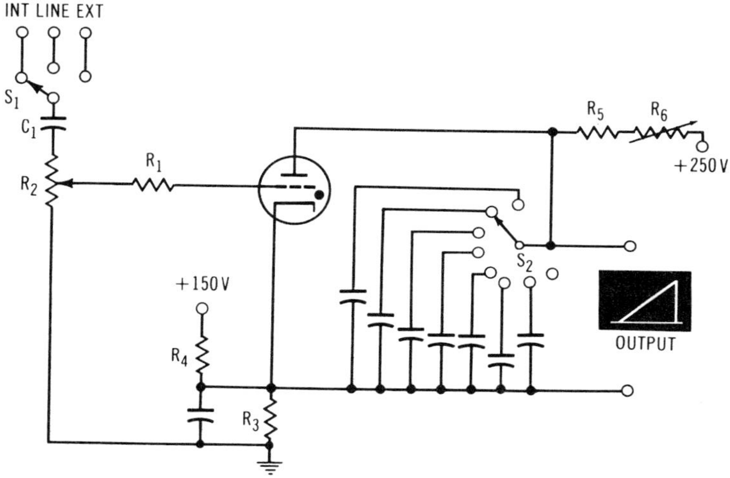

Fig. 4-54. Thyratron sawtooth generator.

which is made up of resistors R_3 and R_4. The grid is thereby maintained at a desired negative bias, since it is grounded through R_1 and R_2. The bank of capacitors across the tube represents the individual coarse settings for sawtooth rise time. The selected capacitor is charged by the B+ source through resistors R_5 and R_6. Resistance R_6 can be adjusted for the precise rise time desired. Because of its established ionizing and deionizing potentials, the thyratron acts like a stable rapid switch in charging and discharging the chosen capacitor. The sawtooth waveforms developed across the capacitor are fed to the next stage, the horizontal amplifier.

Q4-67. What is meant by the deionizing potential of a thyratron?

Q4-68. The cathode of the thyratron is kept at a _ _ _ potential.

Q4-69. What factors make the thyratron useful in a sawtooth-generating circuit?

Your Answers Should Be:

A4-67. The plate voltage at which the thyratron **stops conducting** is known as the deionizing potential.

A4-68. The cathode of the thyratron is kept at a **low** potential.

A4-69. **Ionization** and **deionization.**

Solid-State Sawtooth Generator

In solid-state oscilloscopes, a transistor multivibrator is often used as the sawtooth generator. You were introduced to the multivibrator in Volume 3 of this series of books. An asymmetrical type of multivibrator is used for a sawtooth generator. The circuit in its basic form is shown in Fig. 4-55.

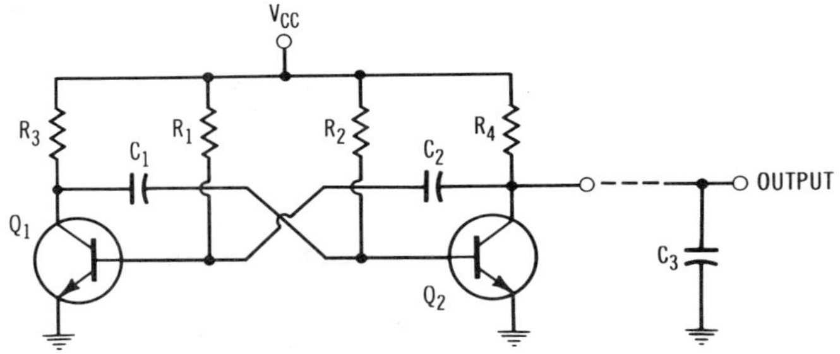

Fig. 4-55. An asymmetrical multivibrator.

A multivibrator may be thought of as two resistance-coupled amplifiers arranged so that the output of each amplifier is fed back to the input of the other. If you consider that the circuit of transistor Q_1 is such an amplifier, you will see that its output is coupled through capacitor C_1 back to the base of transistor Q_2. Also, the output of C_2 is coupled through C_2 back to the base of transistor Q_1.

How does the multivibrator start its operation? When the circuit is first supplied with power, one of the transistors—let's say, Q_1—starts to conduct sooner than the other. Thus, Q_1 is "on" and transistor Q_2 is "off." Current through Q_1 increases to the saturation point. Then, transistor Q_2 starts to conduct and its current increases while the current of Q_1 decreases—until Q_2 reaches saturation (transistor Q_2 is on). Meanwhile, the current of transistor Q_1 stops—Q_1 is off. Then, the process reverses. The conduction of Q_2 stops increasing and that of Q_1 starts to increase. This action happens over and over, and a stream of sawtooth waves is generated.

If $C_1 = C_2$ and $R_1 = R_2$, a symmetrical waveform will be produced as shown in Fig. 4-56A. If capacitor C_2 is much smaller than C_1, the RC time constant for the base circuit of transistor Q_2 will be shorter than the time constant for transistor Q_1. The result is that the multivibrator action of transistor Q_2 will have a short negative-going pulse and a longer positive-going pulse, as shown in Fig. 4-56B.

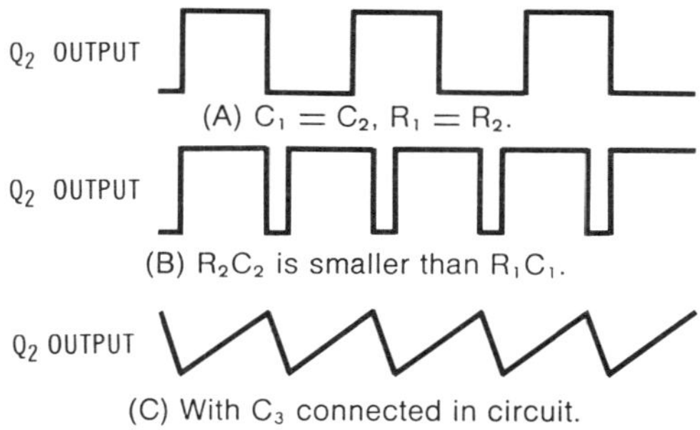

Fig. 4-56. Output waveforms for circuit in Fig. 4-55.

Capacitor C_3 is used to shape the output pulse (Fig. 4-56B) of the multivibrator into a sawtooth waveform as shown in Fig. 4-56C. During the positive portion of the pulse (Fig. 4-56B), capacitor C_3 charges providing the rising portion of the sawtooth waveform. Then, it quickly discharges as the leading edge of the negative pulse arrives. With the arrival of the trailing edge of the negative pulse, capacitor C_3 begins charging again to restart the sawtooth-forming process.

Q4-70. For generating sawtooth waveshapes, an _ _ _ - _ _ _ _ _ _ _ _ _ _ multivibrator is often used.

Q4-71. A solid-state multivibrator starts to conduct when one of the _ _ _ _ _ _ _ _ _ _ _ _ starts to conduct.

Q4-72. In the multivibrator circuit of Fig. 4-55, if $C_1 = C_2$ and $R_1 = R_2$, a _ _ _ _ _ _ _ _ _ _ _ _ _ waveshape is produced.

Q4-73. The purpose of capacitor C_3 is to shape the output pulse into a _ _ _ _ _ _ _ _ _ waveform.

> **Your Answers Should Be:**
>
> **A4-70.** For generating sawtooth waveshapes, an **asymmetrical** multivibrator is often used.
>
> **A4-71.** A solid-state multivibrator starts to conduct when one of the **transistors** starts to conduct.
>
> **A4-72.** In the circuit of Fig. 4-55, if $C_1 = C_2$ and $R_1 = R_2$, a **symmetrical** waveshape is produced.
>
> **A4-73.** The purpose of capacitor C_3 is to shape the output pulse into a **sawtooth** waveform.

Controlling Frequency and Timing of the Sawtooth

Two controls for the **sweep-oscillator** circuit are on the front panel of the scope. The COARSE FREQUENCY control selects one of seven capacitors in the circuit of Fig. 4-54. Numbers on the switch specify the frequency of the sawtooth waveshape.

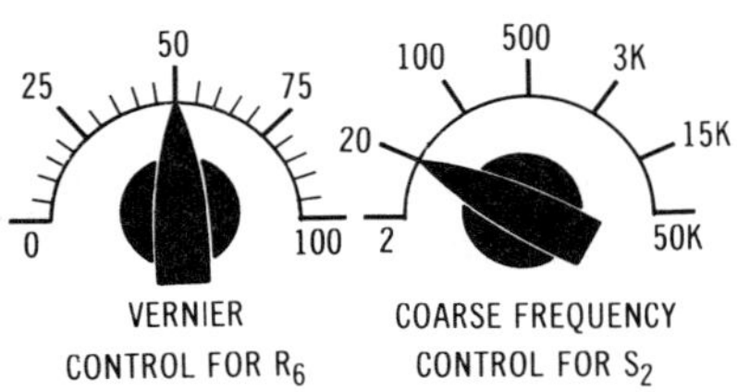

Fig. 4-57. Sweep-oscillator controls.

The VERNIER control (Fig. 4-57) makes a fine setting of resistance R_6 to obtain frequencies between the coarse settings. To place a 60-Hz waveform on the screen, for example, the COARSE FREQUENCY control is set on 100 and the VERNIER control is adjusted until a single cycle is presented.

Sync Circuits

You may have noted the 3-position switch (S_1) shown in the thyratron circuit of Fig. 4-54. That part of the circuit is shown again in Fig. 4-58. The purpose of a sync circuit is to cause the sawtooth waveform to remain in **synchronization** with the waveform that is to be placed on the screen. That is, both waveforms must start at the same time. The origin of the waveform to which the sawtooth is to be synchronized determines the setting of the SYNC SEL (sync-selector) switch

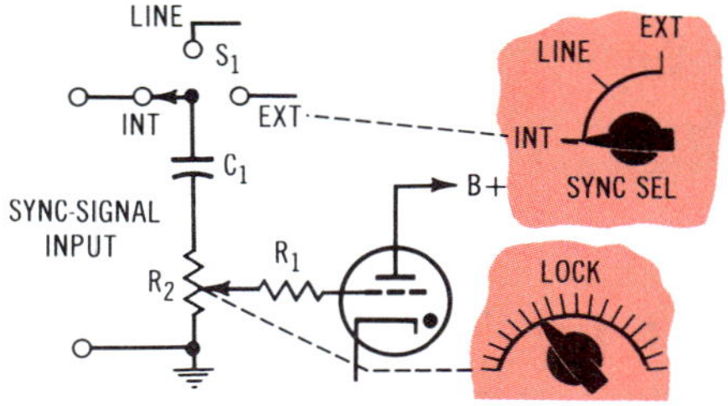

Fig. 4-58. Sync circuit.

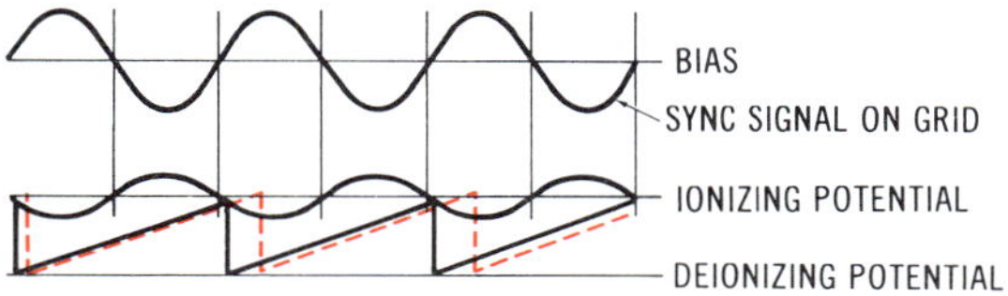

Fig. 4-59. Synchronizing sequence.

on the front panel. **EXT** (external) is the setting used when the sync signal is obtained from an external circuit or source. **LINE** obtains the sync signal from the scope power line. **INT** (internal) samples the waveform in the vertical-deflection amplifier channel. The principle is identical for all three settings. Assume the switch is set on INT. The waveform (on the screen) is fed through C_1 and R_2 (RC coupling circuit in Fig. 4-58). The grid voltage will rise and fall with the amplitude of the signal, thereby decreasing and increasing the time interval before the tube ionizes and conducts current.

A sync signal on the grid will cause a fall and rise in ionization potential, as shown in Fig. 4-59. Without the sync signal, the ionization potential is steady, and the sawtooth waveform is as shown by the dashed line in Fig. 4-59. When the sync voltage is added, the sawtooth voltage reaches the ionization potential sooner in each cycle. The rise time of the sawtooth waveform is shortened, and its frequency is increased. This is shown by the solid waveform.

The LOCK control varies the amplitude of the signal appearing on the grid. This control is necessary since sync signals vary widely in amplitude. A steady uniform sync signal can be obtained by adjusting for proper ionization variation with the LOCK control. Sawtooth generators in solid-state instruments use a similar synchronization action.

Q4-74. Explain the three SYNC SEL switch settings.

Your Answer Should Be:

A4-74. The three settings are **INT (samples the internal signal** in the vertical-deflection circuit), **EXT** (used when the **sync signal is obtained from an external source), and LINE** (used when **sync signal is obtained from the scope power line**).

Horizontal Channels

The sync circuit, sweep oscillator, and horizontal-deflection amplifier make up the horizontal channel.

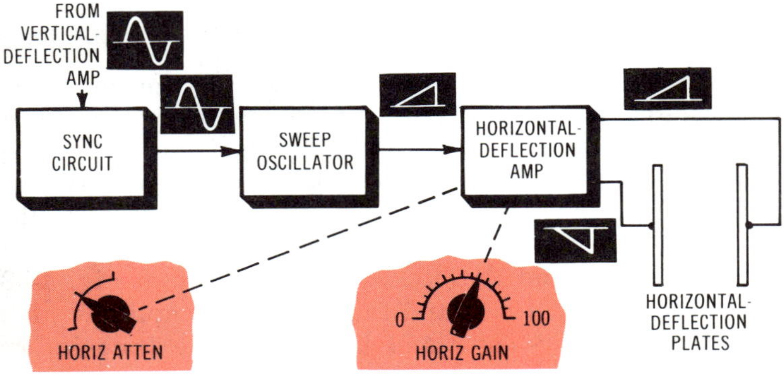

Fig. 4-60. Horizontal channel.

The **sync circuit** sends a sample of the observed waveform to the **sweep oscillator** for synchronization with the generated sawtooth waveshape. The sawtooth waveform is then amplified by the **horizontal-deflection amplifier** and applied (in opposite polarity) to each of the horizontal plates.

Vertical and horizontal amplifiers are similar and perform identical functions. Each has a gain control to develop the desired size of the pattern. Each also has an attenuation control to decrease the amplitude of large waveforms so that they will be retained within the area of the screen. The **HOR ATTEN** control is used when an external waveform is to be applied to the horizontal-deflection plates through the amplifier. Ac and ground jacks are available on the front panel for this purpose.

When a waveform is to be applied directly to the horizontal-deflection plates, the sweep oscillator is disconnected from the horizontal amplifier and neither is used for the oscilloscope display.

The Whole Oscilloscope

You have studied the circuits of a typical general-purpose oscilloscope and have learned that they were not difficult to understand, if you were able to recall the fundamentals that you studied in preceding volumes. You are now ready to combine all of these circuits into a complete unit. This combination of circuits makes up the whole oscilloscope. You will learn how to adjust the numerous controls on the front panel and how they influence the pattern on the screen. Recall now the purpose of some of the oscilloscope controls that are shown in Fig. 4-61.

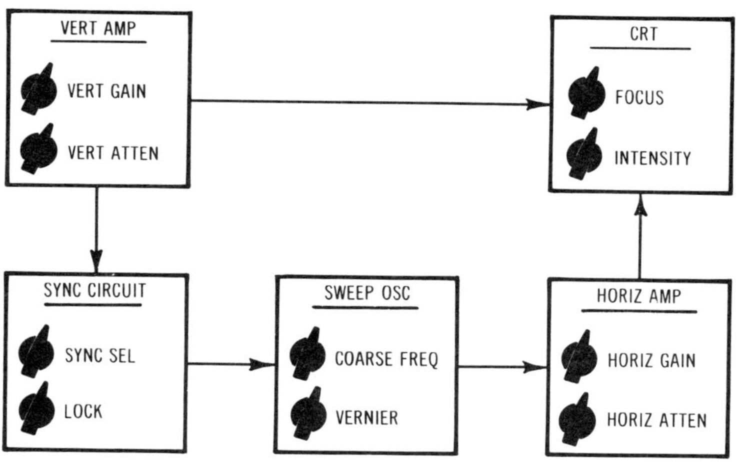

Fig. 4-61. Oscilloscope controls.

Q4-75. State the purpose and how it is accomplished for each of the following controls:

1. Focus control
2. Intensity control
3. Coarse-frequency control
4. Vertical-attenuation control

Your Answers Should Be:

A4-75. The purpose and operation of the various controls are as follows:

1. The focus control **establishes the correct potential differences** between the grid and first anode and between the first and second anode **for focusing.**
2. The intensity control **varies the brightness of the beam** displayed on the screen by controlling the negative bias on the grid.
3. The coarse-frequency control **determines the basic frequency of the sawtooth waveform** by selecting the proper value of capacitance for the RC charging circuit.
4. The vertical-attenuation control **decreases the amplitude of large waveforms** that might be amplified off the screen. This is accomplished by selecting the correct ratio of the waveform voltage from a voltage divider.

Similarity Among Oscilloscopes

The scope you use may differ in some respects from the general-purpose one you just studied. Your unit might be a more advanced scope, such as a dual-trace instrument. Your controls may be identified by different names, and many of the circuits may be designed differently. However, all of the functions will be basically the same. Before using a scope, it is wise to carefully study the manual that comes with it. Descriptions may not be in detail, but the information you have learned so far will help fill in the missing points. Develop a habit of taking all readings with the greatest possible accuracy.

USING THE OSCILLOSCOPE

An oscilloscope can be used for several different types of measurements. Earlier in the chapter, you learned it was most often used to study the shape of a waveform when checking the performance of equipment. The pattern on the scope is compared with the signal that should appear at a test point, and a judgment is then made as to whether the operation of the equipment is good or bad.

You were introduced to the graticule, a plastic sheet scored with calibrated horizontal and vertical lines. By recording the height of a known voltage on the graticule, you can estimate the value of an unknown voltage placed on the screen. Other uses for which an oscilloscope can be used include determining phase relationships and measuring frequencies, as shown in Fig. 4-62. These will be explained later.

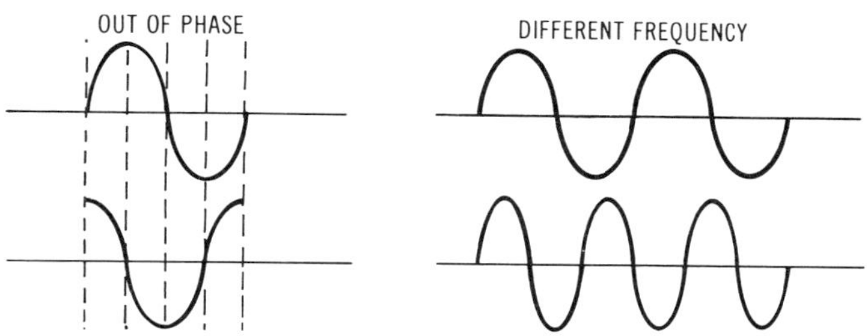

Fig. 4-62. Oscilloscope applications.

Turning the Scope On

First, make sure the scope is plugged into an electrical outlet. Many people turn all the knobs on the front panel out of adjustment before they notice that the power cord is not plugged in. On most scopes, the power switch is part of the INTENSITY control. Turn the knob until a click is heard or a panel light lights. Let the scope warm up for a few minutes so that all the circuit voltages become stabilized.

Getting a Pattern on the Screen

When putting a pattern on the screen, adjust the INTENSITY and FOCUS controls for a bright, sharp line. If other control settings are such that a dot instead of a line appears, turn down the intensity to prevent burning a hole in the screen coating. Brightness and sharpness will vary at various frequency settings, because of the different speeds at which the beam travels across the screen. For this reason, it may be necessary to adjust the INTENSITY and FOCUS controls occasionally while taking readings.

Q4-76. What should you do before you turn on the scope?

Your Answer Should Be:

A4-76. You should carefully **study the manual** that comes with the scope before turning the scope on.

Number of Cycles on the Screen

Because distortion may exist at the beginning and end of a sweep, it is best to put two or three cycles of the waveform on the screen instead of only one (Fig. 4-63). The center cycle

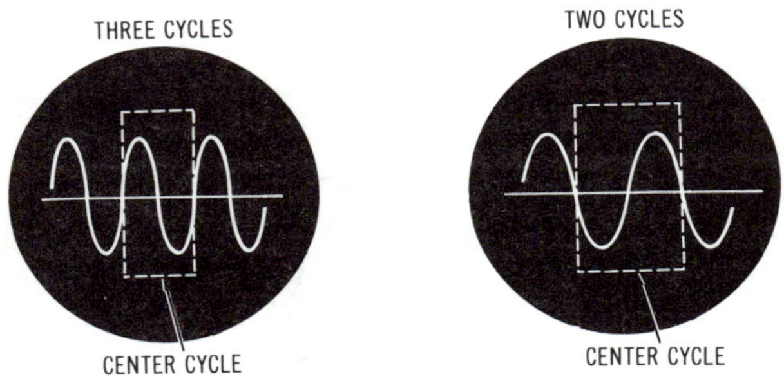

Fig. 4-63. An undistorted waveform.

of three cycles gives you an undistorted waveform in its correct phase. The center of a two-cycle presentation will appear inverted, but will be undistorted.

The relationship between the frequencies of the waveform on the vertical plates and the sawtooth on the horizontal plates

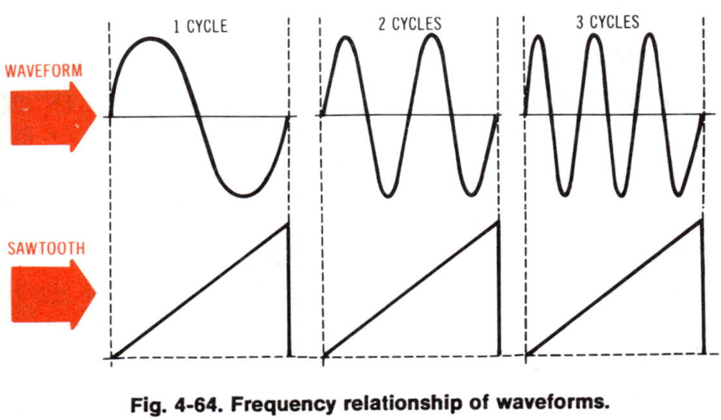

Fig. 4-64. Frequency relationship of waveforms.

determines the number of cycles on the screen (Fig. 4-64). The sweep frequency should always be kept lower than or equal to the waveform frequency; it should never be higher. If the sweep frequency were higher, only a portion of the waveform would be presented on the scope.

As Fig. 4-64 shows, three cycles of the waveform will be on the screen when the sweep frequency is set to ⅓ the frequency of the input signal. If the input frequency is 12,000 Hz, the sweep frequency must be 4000 Hz for a three-cycle scope presentation. For two cycles, the sweep frequency must be set at 6000 Hz. If a single cycle is desired, the setting is the same as the input frequency, i.e., 12,000 Hz.

The sawtooth frequency is selected by settings of the COARSE FREQ and VERNIER controls on the front panel. If the exact frequency number is not found on the coarse-frequency markings, set the coarse control to the closest number and adjust the vernier control for a stationary pattern on the screen.

The ratio of waveform to sawtooth frequencies (Fig. 4-65) should be such that it is on the order of 1/1, 2/1, 3/1, 4/1, etc. When the ratio leaves a quotient that is not a whole number (3/2 for example), the display will be a series of lines moving across the screen. If the pattern appears to be incorrect, adjust the proper control (COARSE or VERNIER).

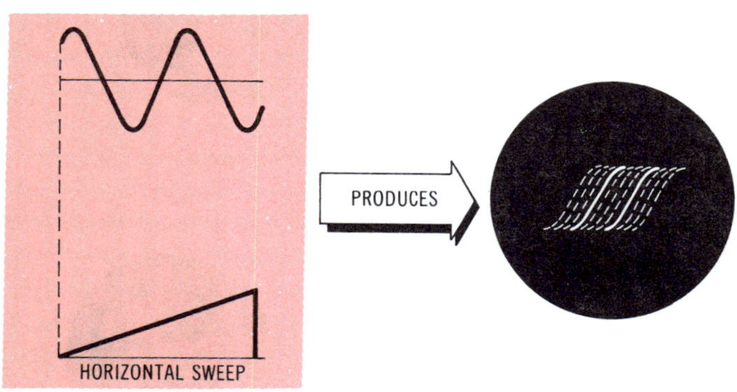

Fig. 4-65. Ratio of waveform to sawtooth frequencies.

Q4-77. Why is there only one sawtooth cycle in each example in Fig. 4-64?

Your Answer Should Be:

A4-77. There can be only one sawtooth wave per sweep of the scope, regardless of the number of waveform cycles. The sawtooth waveform will move the electron beam horizontally across the scope during its rise time. When the sawtooth decays, the beam immediately returns to the starting point on the left.

Other Make-Ready Settings

So far you have learned how to set the INTENSITY and FOCUS controls for proper brightness and sharpness of the waveform on the screen. A steady pattern has been obtained by setting the COARSE FREQ and VERNIER controls of the sweep-oscillator stage for a steady uniform pattern. If the pattern looks like either of the waveforms in Fig. 4-66, you are ready to proceed with an analysis of the waveform.

However, you may obtain patterns that appear similar to those shown in Fig. 4-67.

Fig. 4-66. Uniform steady waveshape patterns.

(A) (B)

Fig. 4-67. Off-center waveshape patterns.

Which control would you adjust for Fig. 4-67A? Which would you adjust for Fig. 4-67B? The waveform in Fig. 4-67A can be adjusted by turning the **VERT POS** control to bring the waveform to the center of the screen. The waveform in Fig. 4-67B can be adjusted by turning the **HOR POS** control to bring the waveform to the center of the screen.

The two positioning controls above and to either side of the screen are used to center the waveform. Suppose the display for two cycles appeared like either one of the pictures in Fig. 4-68. Which controls would you adjust?

(A) (B)

Fig. 4-68. Expanded waveshape patterns.

It is evident that portions of the waveform are being deflected off the screen in both cases. The figure on the left (Fig. 4-68A) has too much horizontal expansion, so you would reduce the HOR GAIN (horizontal gain) setting. Reduction in VERT gain (vertical gain) would bring the waveform on the right (Fig. 4-68B) back on the screen. For normal viewing purposes, the height and width of a waveform pattern should be about equal and should cover about 60 to 70% of the screen. On a 5-inch scope, this would be a little over 3 inches. The pattern should cover about 2 inches for a 3-inch crt. Adjustments of the vertical-gain control not only change the amplitude of the signal fed to the vertical-deflection plates, but they also increase or decrease the amplitude of the signal fed to the sync circuit. Quite frequently, the change will affect the ionizing potential of the thyratron sufficiently to cause distortion in the presentation. To remedy this, adjust the LOCK control in the sync circuit.

Occasionally, a waveform frequency will be encountered that is so high that the frequency of the sweep oscillator cannot be made high enough to give a screen presentation of 2 or 3 cycles. If the upper limit of the COARSE FREQ control were 50 kHz and the frequency of the waveform were 1 MHz, the vertical-to-horizontal ratio would be 1,000,000/50,000 or 20/1. Twenty cycles of the waveform would appear on the screen at this setting. The VERNIER adjustment might eliminate a few cycles, but the remaining cycles would be too close together to permit observation.

When there are too many cycles for easy viewing, expanding the presentation with the horizontal-gain control will separate the cycles for better viewing.

(A) 15–20 cycles. (B) Expanded.

Fig. 4-69. Multicycle presentations.

READING WAVEFORMS

Earlier in this chapter, you learned that signals (waveforms) are modified, or changed, as they pass from circuit to circuit until the signal from the final stage contains the de-

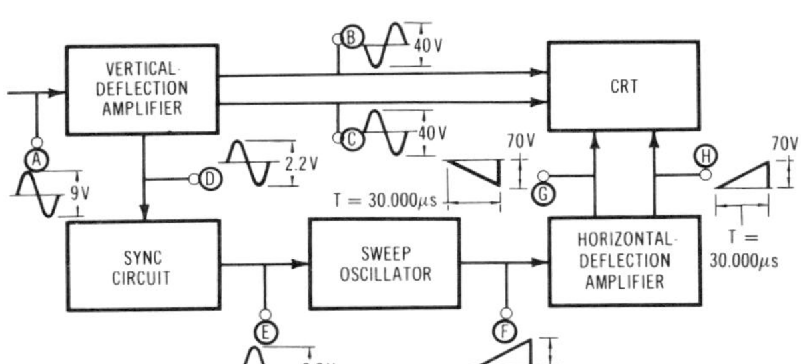

Fig. 4-70. A servicing block diagram.

sired characteristics. With an oscilloscope, you can test each of the signals in a piece of equipment to determine whether the circuits are operating properly and/or which one might be the cause of a trouble.

The illustration in Fig. 4-70 is representative of a **servicing block diagram.** The letters in circles identify significant test points. The waveforms beside them show the shape, voltage, and time duration (where applicable) that should be observed at these points. Arrows on block-connecting lines show the direction of signal flow. This diagram would be used when matching the characteristics of the waveforms to those shown on an oscilloscope. By using a graticule whose lines have been given voltage values in accordance with the amplitude of a known voltage, the voltage at each of the test points could be checked.

LISSAJOUS FIGURES

The phase relationship between two waveforms and the frequency of a signal can be measured on an oscilloscope. Patterns placed on the screen to accomplish this are called **Lissajous figures.** A Lissajous figure is the pattern obtained when ac signals are applied simultaneously to both sets of deflection plates. The following procedures are typical of most oscilloscopes.

Phase Measurement

When you are measuring the phase difference between two signals, one signal is applied to the vertical and the other to the horizontal input. Turn the sweep off so that there will be

Q4-78. To observe three cycles of a 45,000-Hz signal, at what setting(s) would you adjust which controls?

Q4-79. If test points H, G, and F in the diagram of Fig. 4-70 provide faulty indications and test point E does not, in which circuit is the trouble located?

Q4-80. Patterns placed on the screen to show phase relationships are called __________ ________.

Q4-81. Why would you want to enlarge a waveform presentation on an oscilloscope?

Q4-82. What is meant by the output stage in electronic equipment?

Q4-83. How can the phase relationship between two waveforms be measured using the oscilloscope?

Q4-84. When measuring the phase difference of two signals, their frequencies must be _____.

Your Answers Should Be:

A4-78. Set the **COARSE FREQ control to 15 kHz** and adjust the **VERNIER control for a stable display.**

A4-79. The trouble is in the **sweep-oscillator circuit.** This circuit has a good input, but it has a faulty output.

A4-80. Patterns placed on the screen to show phase relationship are called **Lissajous figures.**

A4-81. The presentation is expanded for **better accuracy** when viewing a waveform.

A4-82. The output stage feeds the desired signal **to the load** of the equipment.

A4-83. **Ac signals are simultaneously applied to each set of deflection plates** when comparing their phase.

A4-84. When measuring the phase difference of two signals, their frequencies must be **equal.**

no sawtooth voltage to interfere with the signal in the horizontal channel. For greatest accuracy, the amplitudes of the two signals should be equal. Adjust the gain controls to ob-

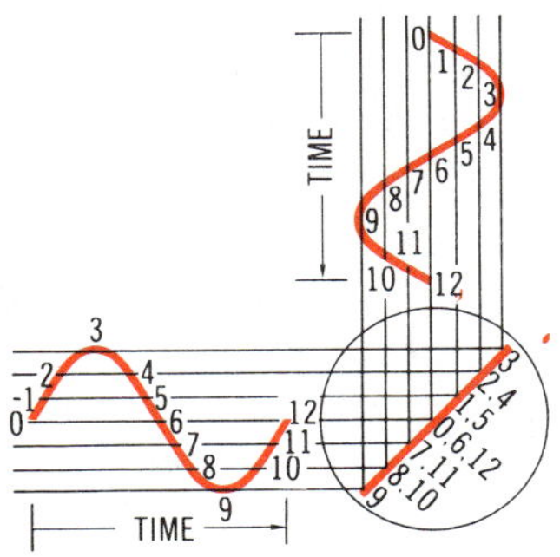

Fig. 4-71. A Lissajous figure—waveforms in phase.

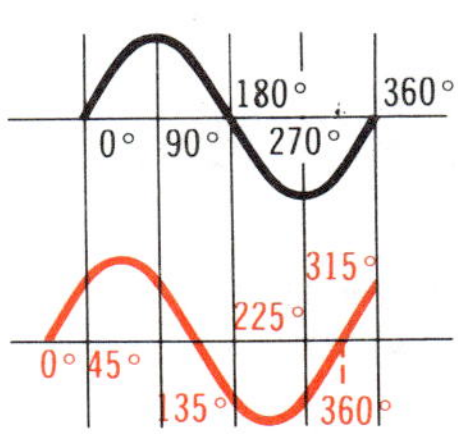

Fig. 4-72. Out of phase waveforms.

tain a pattern that is as high as it is wide. When measuring for phase difference, the two signals must, of course, be the same frequency.

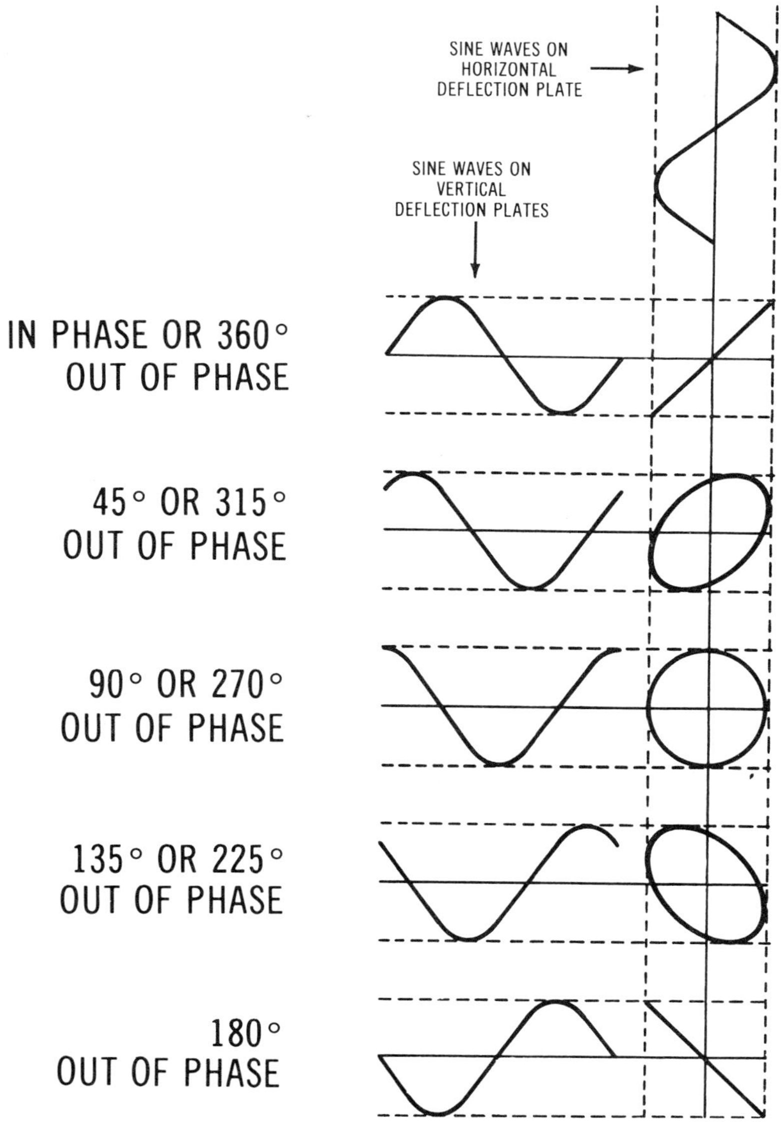

Fig. 4-73. Lissajous patterns—sample phase measurement.

Analyzing a Lissajous Figure

Fig. 4-71 shows a Lissajous pattern for two sine waves. Numbers are assigned to corresponding voltage points on the two signals. Extensions of these points are brought to the screen. The intersection of corresponding numbered lines is the position of the electron beam at that instant of time. In this case, the two sine waves are in phase.

In Fig. 4-72, the voltage/time relationships are different; corresponding voltage points are 45° apart. Therefore, the waveforms are 45° out of phase.

The Lissajous figures shown in Fig. 4-73 are examples of a few out-of-phase relationships. An estimate of the phase difference of two signals can be made by observing the direction and amount of angle and the width of the ellipse. This will be close enough for most checks.

However, if you desire to make the measurement more precisely and can locate values in a sine (trigonometry) table, there is another method available. In this case, the amplitudes of the two signals must be near the same size.

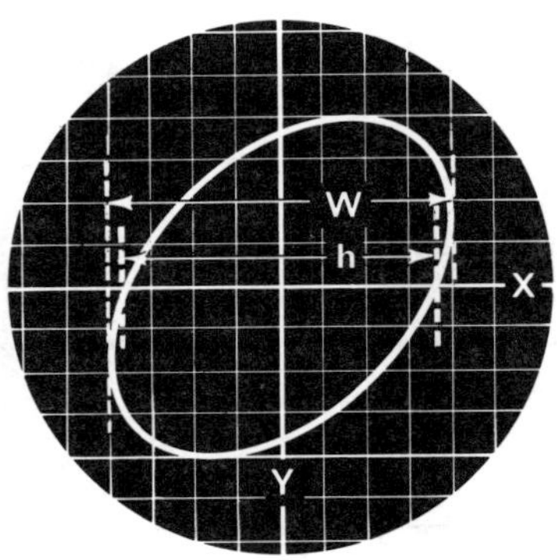

Fig. 4-74. Determining sine of angular phase difference using a Lissajous figure.

The graticule is placed on the crt, and the Lissajous figure is centered as in Fig. 4-74. The overall height and width of the ellipse should be equal in length. The ratio h/W is the sine of the angular phase difference. For example, if h were equal to 7 units and W equal to 8, then:

$$\frac{h}{W} = \frac{7}{8} = 0.875$$

This ratio is approximately the sine of 60°. Therefore, the two signals are 60° or 300° (360° − 60°) out of phase. The distances h and W are shown in Fig. 4-74.

Frequency Measurement

Frequency of an unknown sine wave is determined in a manner similar to phase measurement. A known frequency is applied to the horizontal plates while an unknown waveform appears on the vertical plates. The resulting Lissajous figure will reveal the difference in frequency between the two. The reference frequency could be taken from a calibrated signal generator or from the 60-Hz ac supply. The electron beam will follow the voltage amplitudes placed on the deflection plates. If the waveforms are of the same frequency, the Lissajous figure will resemble phase relationships figures.

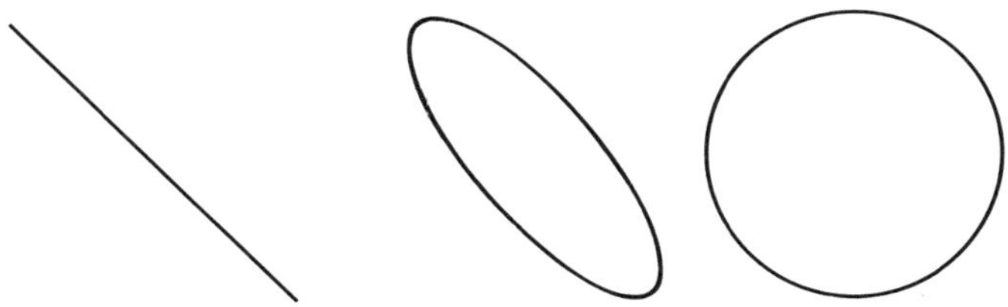

Fig. 4-75. Both frequencies equal.

Frequency relationships can be determined by the number of loops or points that touch the top (or bottom) and one of the sides of the pattern. In the figures so far, there was one

(A) 2/1 ratio. (B) 1/2 ratio.

Fig. 4-76. Tangent points.

Q4-85. What circuits must be bypassed in the horizontal channel to obtain a Lissajous figure on the screen?

Your Answer Should Be:

A4-85. **The sync and sweep-oscillator circuits** must be bypassed. If a sawtooth wave and an external signal were fed to the horizontal-deflection amplifier at the same time, a Lissajous figure could not be developed. The sweep oscillator would be erratic.

point of tangency at the top and one at the side. The frequency ratio was 1/1; the unknown had the same frequency.

In Fig. 4-76A, one cycle of standard frequency appears on the horizontal plates at the same time that two cycles of the unknown are on the vertical plates. There are two points (loops) of top tangency and one point of side tangency. Frequency ratios should be expressed in terms of vertical (unknown) to horizontal (standard). The frequency ratio of Fig. 4-76A is 2/1, and in Fig. 4-76B, it is 1/2.

Additional Samples of Lissajous Figures

Both figures in Fig. 4-77 provide a vertical/horizontal (unknown/standard) ratio of 3/1. But they do not look alike. Fig. 4-77A is known as a **closed** pattern. If you will start at any point in the figure and follow the line, you will return to

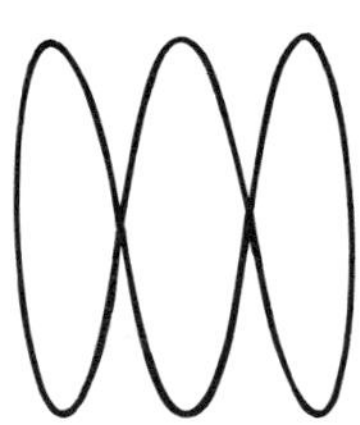

(A) Closed pattern.

(B) Open pattern.

Fig. 4-77. Lissajous ratio of 3/1.

the starting point. The figure is continuous; it has no beginning or end. Fig. 4-77B, however, is not continuous; it has a beginning and an end. Its pattern is **open.**

The three points of tangency at the top and one point of tangency at the side are easy to count in Fig. 4-77A. But the three tangent points do not appear in Fig. 4-77B. The problem is resolved by counting tangency points in halves instead of units in an open pattern. Each line that terminates at the top or side is a one-half point of tangency. Each loop is considered to be two ends or two one-half points. Counting at the top, there is one loop (two halves) plus one end (one half) for a total of three halves. At the side, there is a single end for one half. The vertical/horizontal ratio is three halves divided by one half, or 3/1.

A continually shifting Lissajous pattern results when the phase relationship between the two input signals is constantly changing. The more complex the pattern (resulting from a frequency ratio having large numbers, such as 17/13), the harder it is to interpret. It is better, then, to simplify the ratio by changing the known frequency. Other frequency measurements using Lissajous figures are shown in Fig. 4-78.

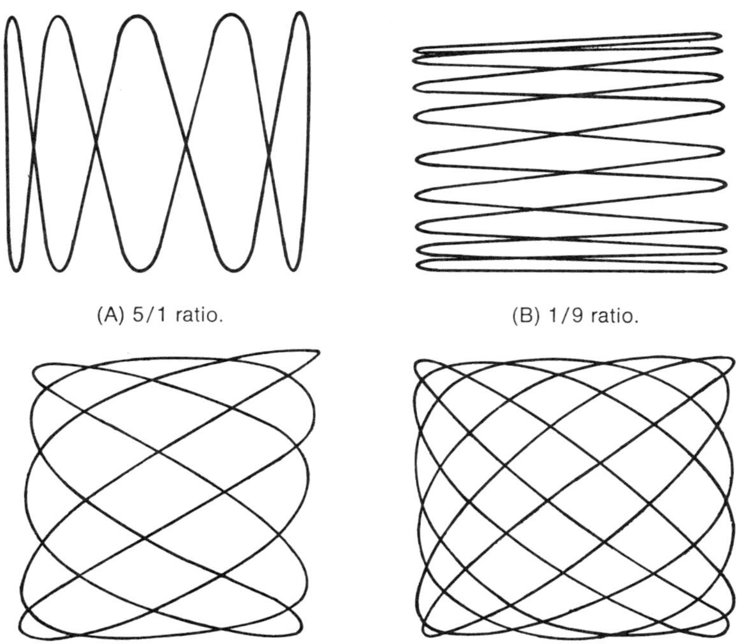

(A) 5/1 ratio.

(B) 1/9 ratio.

(C) 3/5 ratio.

(D) 5/6 ratio.

Fig. 4-78. Lissajous figures.

Q4-86. If the standard frequency is 180 Hz, what is the frequency of the unknown in each of the Lissajous patterns in Fig. 4-78?

Your Answers Should Be:

A4-86. Values of unknown frequencies are:

(A) **900 Hz;** vertical/horizontal ratio is 5/1.

(B) **20 Hz;** 1/9 ratio.

(C) **108 Hz;** 3/5 ratio.

(D) **150 Hz;** 5/6 ratio.

TRIGGERED-SWEEP OSCILLOSCOPES

The kind of oscilloscope that we have been studying up to this point is the **recurrent-sweep** oscilloscope. That is, the sawtooth oscillator runs continuously, generating a sawtooth waveform over and over. The sawtooth oscillator is adjusted to be at the same frequency as the frequency of the input signal—or at a submultiple frequency.

Although the recurrent-sweep oscilloscope is a very useful general-purpose oscilloscope, a requirement for more precise analysis of television signals, pulses, computer waveforms, and other signals, has brought about the development of the **triggered-sweep** oscilloscope. Many oscilloscopes that are now in use are of the triggered-sweep, or triggered time-base, type.

A triggered scope is usually calibrated in terms of **microseconds per centimeter** of horizontal deflection. The graticule covering the face of the crt is marked off in centimeters. With a triggered oscilloscope, it is easily possible to measure dimension characteristics of pulses and other waveforms. For example, measurement of the rise time, decay time, width, interval, and other dimensions of rectangular pulses is fairly easily accomplished. A feature normally included in a triggered-sweep oscilloscope is a delay line as part of the vertical amplifier circuit. Its use can best be understood by referring to Fig. 4-79.

The synchronizing signal, which is derived from the input signal just as in the recurrent-sweep oscilloscope, goes through a trigger-pulse shaper and a sweep generator, before being sent through the horizontal amplifier and, then, to the horizontal-deflection plate of the cathode-ray tube. At the same time, the signal to be observed on the face of the screen goes through a delay line in the vertical-amplifier circuit which delays the signal about 0.5 microsecond before reaching the crt. Without this delay, the horizontal-sweep circuit would start at about the same time that the vertical-channel signal arrived at the crt and some of the leading edge of the wave-

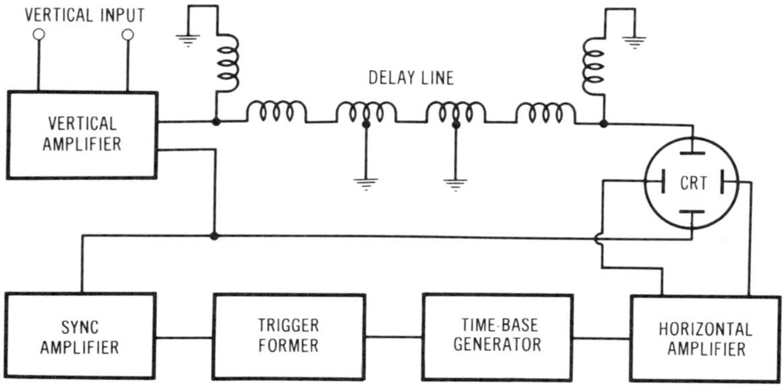

Fig. 4-79. Vertical amplifier circuit using a delay line.

form on the screen would not be seen in the display. The delay line holds back the signal, that is to be displayed, just long enough for the horizontal sweep to get started.

An example of the use of a triggered oscilloscope to measure waveform dimensions is shown in Fig. 4-80.

On the left is a video waveform (Fig. 4-80A), while on the right (Fig. 4-80B), you have a representation of just the horizontal sync and blanking portion of the waveform. The object here is to measure the front-porch part of the blanking pulse (that part just to the left of the sync pulse) and the width of

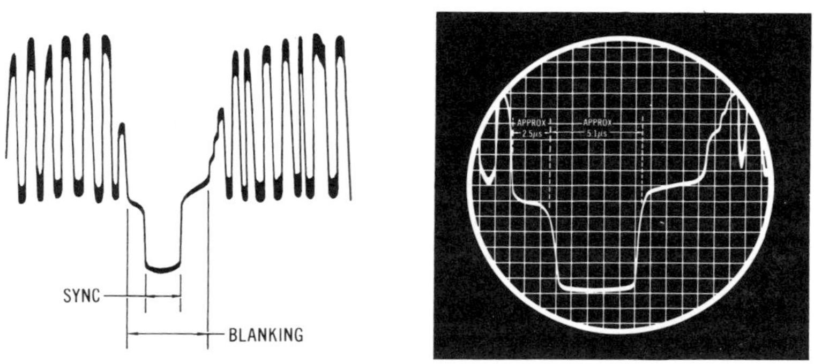

(A) Video waveform.

(B) Expanded sync pulse.

Fig. 4-80. Using a triggered oscilloscope to measure waveform dimensions.

the sync pulse itself. Assuming that the calibrated time base of the oscilloscope is set for 1 microsecond per centimeter, and that each division on the crt face represents one centimeter, you can see that the front porch of the blanking pulse is about 2.5 μs and the sync portion is about 5.1 μs in width.

The pulse could be spread out even farther by setting the calibrated time-base control to a faster scanning rate—say, 0.1 μs per centimeter. Then, the rise time of the sync pulse could also be measured quite accurately.

MULTITRACE OSCILLOSCOPES

A feature often included in a triggered-sweep oscilloscope is a means of viewing two or more signals at the same time, one positioned over the other. With a multiple-trace feature, the input and output signals of an amplifier can be viewed simultaneously, or one waveform may be an expanded portion of the first waveform, or two or more signals that are at different points in a series of counter circuits, can be seen all at the same time.

Most multiple-trace oscilloscopes are dual-trace instruments. That is, they can show only two simultaneous signals at a time as in Fig. 4-81. The upper trace represents the input to a 4-to-1 counter. Its output is represented by the lower trace.

There are two main methods of obtaining dual oscilloscope traces. One is the **alternate-channel mode** of display. The other is the **chopped-mode** type of display. Both methods can be represented by Fig. 4-82. The block diagram shows two separate input channels. The advantage of the alternate-channel mode of display is a continuous waveform display, even though each

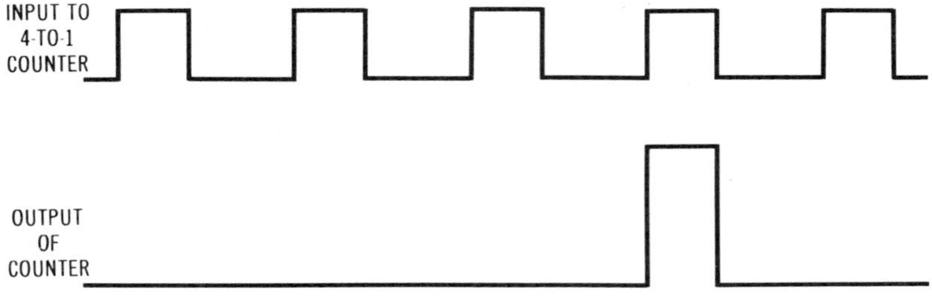

Fig. 4-81. Dual-trace or simultaneous waveforms.

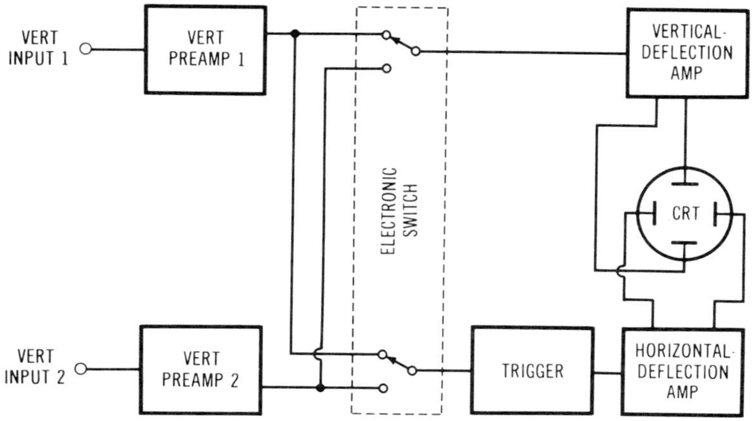

Fig. 4-82. Block diagram for a dual-trace oscilloscope.

trace is displayed only one-half the time. Persistence of vision makes the displays seem to be continuous except at very low scanning frequencies, at which time flicker becomes objectionable. At that point, a satisfactory display can be obtained by changing over to the chopped-mode type of operation. The chopped-mode type of display can operate at low scanning rates without any noticeable flickering.

For both modes, you should understand that it is necessary to provide alternate positive and negative dc voltages to the signals that are applied to the crt vertical plates so that the alternate waveforms will be above and below each other. Most dual-trace scopes provide a switch to select between alternate

Q4-87. To ensure that the signal that is to be seen on the crt does not arrive at the cathode-ray tube before the sawtooth deflection wave arrives, a _ _ _ _ _ line is used in the vertical amplifier.

Q4-88. A triggered-sweep scope is generally calibrated in _ _ _ _ _ _ _ _ _ _ _ _ _ _ per centimeter.

Q4-89. If you wish to view three waveforms in different parts of a circuit simultaneously, you would use a _ _ _ _ _ _ _ _ _ _ _ oscilloscope.

Q4-90. The two multitrace oscilloscope operating modes are the _ _ _ _ _ _ _ _ _ _ _ _ _ _ _ _ _ _ mode and the _ _ _ _ _ _ _ _ mode of operation.

Your Answers Should Be:

A4-87. To ensure that the signal to be seen on the crt does not arrive at the cathode-ray tube before the saw-tooth deflection wave, a **delay** line is used in the vertical amplifier.

A4-88. A triggered-sweep scope is generally calibrated in **microseconds** per centimeter.

A4-89. If you wish to view three waveforms in different parts of a circuit simultaneously, you would use a **multiscan** oscilloscope.

A4-90. The two kinds of multi-trace oscilloscope operating modes are the **alternate-channel** mode and the **chopped** mode of operation.

and chopped modes of operation. A more expensive dual-trace scope is the dual-beam scope which uses a crt with two separate guns.

STORAGE OSCILLOSCOPES

A storage oscilloscope can display and retain a waveform on the face of the crt for a prolonged examination. In one type of storage oscilloscope, the phosphor globules are insulated from one another on the faceplate and, once lit, they stay illuminated by a stream of electrons until erased. In another type, two wire meshes are used behind the phosphor. A charge on one mesh prevents electrons from passing through. The electron gun overcomes the charge and forces electrons through the mesh to produce a trace.

WHAT YOU HAVE LEARNED

1. You obtained additional practice in analyzing the way in which circuits operate.
2. You also gained experience in reading schematic and block diagrams by tracing signals (waveforms) from stage to stage through the diagrams. A servicing block diagram was introduced. Waveform information contained in such diagrams was found to be useful in checking circuit operation.

3. An oscilloscope is a test instrument capable of presenting waveforms. An oscilloscope reproduces the amplitude and time characteristics of a waveform. You can use this capability to check the condition of waveforms at selected test points in many kinds of electronic equipment. The information you obtain will tell you the operating status of a circuit or help you to isolate trouble to a single circuit.
4. A scope can be used for other tests in addition to checking the shape of a waveform. Voltage measuring is one test. Since an oscilloscope has a relatively uniform deflection sensitivity (inches per volt) or sensitivity factor (volts per inch) across its screen, this feature can be used to estimate the peak-to-peak voltage of a waveform placed on the screen. A graticule (plastic sheet containing horizontal and vertical lines) and a voltage standard are required. The graticule is placed on the face of the scope, and the standard voltage is applied to the vertical plates.
5. Phase measurement is another test. The standard signal is applied to the horizontal amplifier, and the unknown is applied to the vertical amplifier. The resulting Lissajous pattern determines the phase relationship of the two signals, when frequency and amplitude are equal.
6. Lissajous patterns can also be used in measuring frequency. A known frequency is applied to the horizontal amplifier; the unknown frequency is applied to the vertical amplifier. By counting the number of tangency points at the top and at one side, a ratio of the unknown to the known frequency can be obtained. After multiplying the ratio times the known frequency, you have the frequency of the unknown quantity.
7. An oscilloscope contains two basic sections—the crt and control circuitry. The crt is designed to place a controllable beam of electrons on the face of the tube. The circuitry controls the movement of the beam.
8. An electron gun contains a cathode (to emit electrons), a control grid (to control the intensity of the trace on the screen), a first anode (to develop the electric lenses that focus the beam on the screen), and a second anode

(to accelerate the electrons toward the screen). Deflection plates in vertical and horizontal pairs are used to position the beam on the screen. If a waveform is applied to the scope, the plates deflect the beam according to the amplitude and time characteristics of the waveform. The screen is made of fluorescent materials that give off light when struck by fast-moving electrons. The picture seen on the screen is formed by the illumination of these materials.

9. The control circuitry has two channels—vertical and horizontal. A constant-gain amplifier places the waveform to be measured on the vertical-deflection plates of the crt. The beam follows the differences of potential between the two plates and, therefore, the amplitude of the waveform.
10. The horizontal channel contains a sync circuit, a sweep oscillator, and an amplifier similar to that used in the vertical channel. The sync circuit obtains a synchronizing signal from the vertical amplifier, the power line of the scope, or from an external source. The sync signal is applied to the sweep oscillator to synchronize its frequency in phase with the waveform in the vertical channel. The shape of the sawtooth waveform is such that when it is amplified, it will be the precise time base required to place one, two, three, or more waveforms on the screen at one time. The horizontal channel can be used for bringing an external signal into the scope.
11. Intensity and focus controls vary the brightness and sharpness of the picture.
12. There are two sync-circuit controls. The sync-selector switch is used for selecting the correct sync signal. A lock control stabilizes the screen presentation.
13. To obtain the correct time base for a wide selection of input frequencies, a coarse-frequency switch and a vernier control are used. The coarse-frequency switch selects the approximate frequency setting; the vernier makes fine adjustments to obtain a stable waveform.
14. Study the manual that accompanies an oscilloscope before using it. The manual will provide the information necessary to operate and use the scope properly.

5

Vacuum-Tube and Semiconductor Testers

what you will learn

On completion of this chapter, you will be able to use a tube tester, within its capabilities and limitations, to check the quality of a vacuum tube. You will learn how to test tubes for emission, mutual conductance, shorts, noise, and gas. You will be able to explain how these tests are conducted in a tube tester, and, thereby, judge the validity of the readings. You will also learn how typical semiconductor tests can be made. If you study this chapter thoroughly, you will be able to make simple tests on tubes and semiconductors to determine their operating quality, without the use of special instruments.

Fig. 5-1. The ultimate question.

TUBE-TESTER APPLICATIONS

Many technicians claim that in equipment using vacuum tubes, 50 to 80% of all circuit troubles are caused by bad tubes. Since there are frequently six or seven components in a vacuum-tube circuit, you might think that the component to be tested first would be the tube. Even the lower percentage figure suggests that the law of averages is on your side. Although it is true that vacuum tubes fail more frequently than most other electronic components, an experienced technician would not grab at this statistical fact as the sole reason for putting a tube into a tube tester. As you will discover in this chapter, there are many reasons why tubes fail, and a tube tester will not always reveal all of them.

A tube which checks out as good on the tester may not function correctly in a particular circuit. A tube may appear bad during the tube test and still perform its function in the circuit. Also, a tube may have gone bad as the result of a faulty resistor, capacitor, coil, connector, or switch. For example, if a cathode capacitor shorts, the resulting increase in plate current could damage the tube. In this case, replacing the bad tube with a good one will not correct the trouble. Sooner or later, excessive current will damage the new tube.

These and other limitations of a tube tester, as well as its capabilities, will be described. The point to remember while studying and using a tube tester is the need for common sense and technical judgment in interpreting the readings it may give. Reserve your conclusion that a tube must be bad until it has been technically proven bad.

VACUUM-TUBE CHARACTERISTICS

As in previous chapters, some underlying fundamentals will be reviewed before the test instrument itself is explained. For a tube tester, the fundamentals are the basic characteristics of a vacuum tube.

Tube Types

A vacuum tube consists of several elements inserted in a glass or metal container that has been evacuated (most of the air removed). Electrons move more readily in a vacuum than in air. An exception to the literal meaning of the term **vacuum tube** is a gas tube in which air has been replaced by an ionizing gas that supports electron flow.

All vacuum tubes contain elements that aid or control the flow of current through the tube. A tube may have a heater, cathode, plate, and one or more grids. The combinations determine the type of tube and how it can be used in a circuit. If the circuit is to act as a rectifier or detector, for example, a diode would be used. An amplifier or oscillator circuit would require a triode, tetrode, pentode, or other multielement tube. A circuit that detects and amplifies might use a duodiode triode.

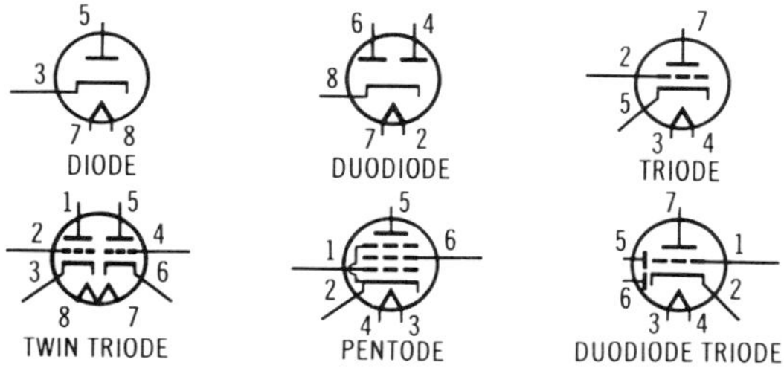

Fig. 5-2. Typical combinations of tube elements.

The tubes shown in Fig. 5-2 are just a few of many combinations of tube elements. To conserve space, equipment manufacturers use multipurpose tubes as much as possible. The duodiode, twin triode, and duodiode triode are just three of many examples. A number next to an element indicates the number of the pin to which it is connected. This is normal practice in most schematic diagrams. Several different types of tube sockets are required to accommodate the number and spacing of pins in the tube base. Some examples are shown in Fig. 5-3.

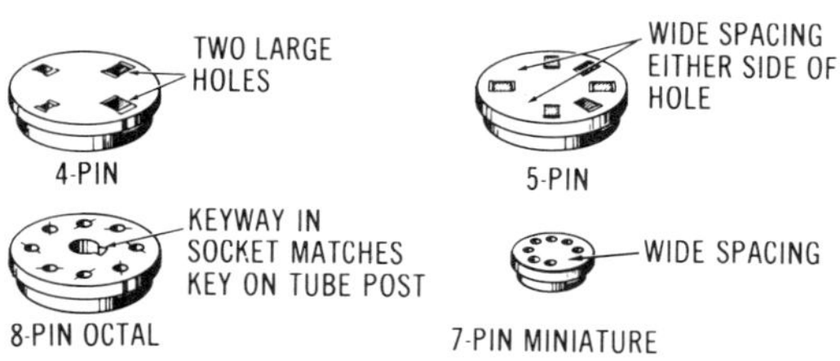

Fig. 5-3. Typical tube sockets.

Q5-1. When numbering tube pins and socket holes for an 8-pin tube, is the same number always assigned for each element?

Q5-2. Electrons move more readily in a ______ than in ____.

Q5-3. What are the names of the elements that may be contained in a vacuum tube?

Your Answers Should Be:

A5-1. **No.** There is no standard numbering system for tube elements.

A5-2. Electrons move more readily in a **vacuum** than in **air.**

A5-3. A **heater, cathode, plate,** and one or more **grids** may be contained in a vacuum tube.

Tube Defects

Tube elements are mounted very close to one another. Periodic heating and cooling of the tube can cause the metal in the elements to weaken and bend. If there is sufficient bending, neighboring elements could touch and cause a condition of arcing or shorting. To restore proper circuit operation, such a tube must be replaced.

Modern manufacturing techniques are capable of producing a tube that is ruggedly constructed. Under normal operations, a heating element may operate for 2000 hours before it weakens and opens. Its life is shorter under abnormal conditions, such as operating the heater at too high a temperature or operating the tube continuously at its maximum ratings. When the heater or any other element opens, the tube will no longer conduct current properly.

A decrease in the emitting capability of the cathode is another tube defect that frequently occurs. The cathode, when heated, will emit electrons into an area around it. When the plate becomes positive with respect to the cathode, the plate will attract a quantity of electrons. The amount will depend on the difference in potential between the two elements.

Fig. 5-5A shows how an electron cloud surrounds the heated cathode. Fig. 5-5B represents normal current flow between

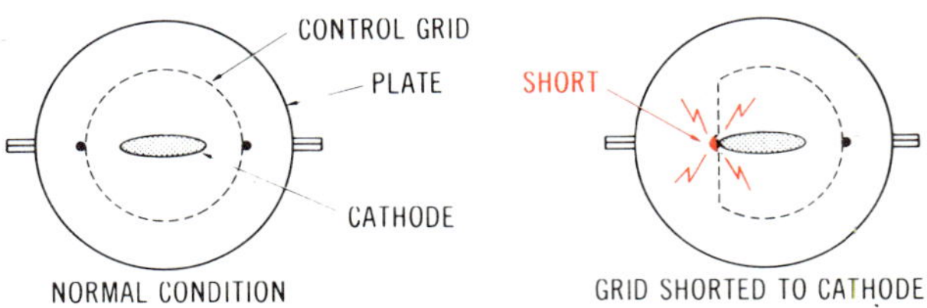

Fig. 5-4. Top view of tube elements.

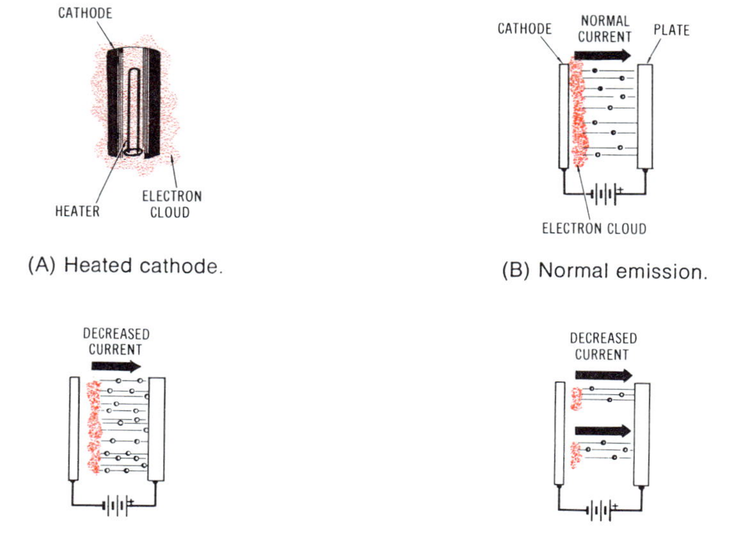

(A) Heated cathode.

(B) Normal emission.

(C) Weakened emission.

(D) Nonuniform emission.

Fig. 5-5. Electron emission.

cathode and plate. In Fig. 5-5C, the emitting capability of the cathode has decreased after many hours of operation. A smaller quantity of electrons in the cloud causes a reduction in plate current. In Fig. 5-5D, the cathode has weakened more in some areas than in others, resulting in a reduced number of electrons available for plate current.

There are other tube defects that occur quite frequently. **Leakage** between the cathode and heater occurs when electrons travel from the cathode to the heater after the two have become shorted or partially shorted. This leakage flow reduces the quantity of electrons available for full plate current. If leakage is intermittent, it could add noise to the circuit.

Vibration or overheating may loosen the elements or their supporting wires, causing **microphonics.** The tube picks up vibrations and acts like a microphone. This happens when the vibrations cause changes in the capacitance that normally exists between tube elements. Tubes can become **gassy** when gas trapped in the metal of the tube elements is released. This gas interferes with the normal operation of the tube.

Q5-4. List some of the common tube defects.

> **Your Answer Should Be:**
>
> A5-4. Some of the common tube defects are: **shorting, arcing, open element, decrease of cathode-emitting capability, leakage current flow, loose elements, and gas inside the envelope.**

Defect Check

Many times a defective tube can be located without using a tube tester. Some of these checks are described below.

Sight Check—If the tube is in the equipment with voltage applied, there are visual checks you can make. Look down toward the base of the tube. A small red glow will, in most cases, indicate that the heater is operating.

If one heater in a series string opens, none of the other heaters will receive current. If you are using a sight check, look for the red glow in neighboring tubes. If the glow is missing, a heater may be open in any one of the tubes in the string. Fig. 5-6A shows the way in which a series string of heaters is shown on a schematic diagram. From the schematic, you can also identify the pin numbers of the tube. With an ohmmeter, you can determine whether or not the heater has opened. With the tube out of the socket, place the probes of the meter on the heater pins. If the heater is open, the pointer will not deflect (infinite resistance).

Also check the plates of the tubes. If any are glowing red, the tube could be drawing excessive current. If this is so, the

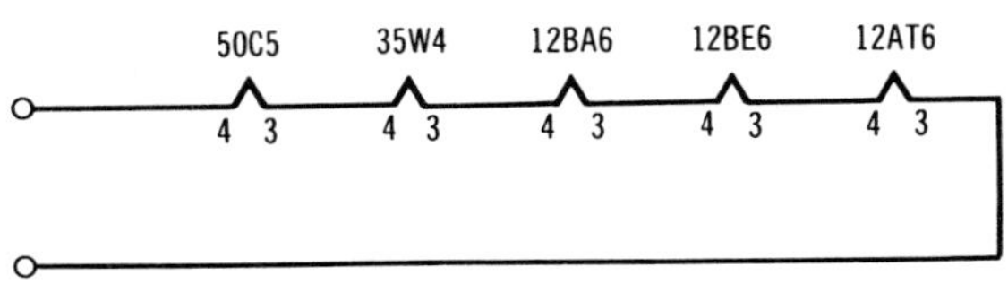

(A) Heaters in series.

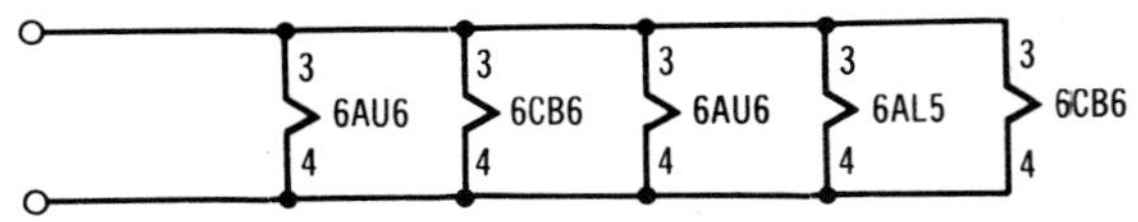

(B) Heaters in parallel.

Fig. 5-6. Connection of vacuum-tube heaters.

trouble could be in the tube itself or in its circuit. Some tubes are designed to operate at a red-hot temperature; these are usually found only in circuits that operate at periodic intervals instead of continuously.

Another visual check can be made. A blue glow inside the envelope indicates a gassy tube. Gas molecules in the tube are being bombarded by plate-current electrons and some of the energy is being released in the form of a blue light. Of course, it is normal for tubes that contain an ionizing gas to glow when they operate. A sight check for gas may not always be conclusive. There may not be enough molecules to cause a glow, but there may be a sufficient number to disrupt the plate current.

Touch Checks—Most tubes are warm or hot when plate current is flowing. Touch the tubes only momentarily, since many tubes operate at extremely high temperatures. A hot metal or glass tube usually indicates that the heater is operating.

Tubes with loose elements can be detected when they are being used in a radio receiver or other equipment having an audible output. Sharply tap the tube and listen for noise (called microphonics) in the speaker.

Substitution Test—Substituting a tube known to be good for one suspected of being bad is another test that can be made without a tube tester. Try to isolate the faulty circuit or group of circuits, and then substitute one tube at a time, listening or watching for a change in equipment operation. If no improvement is noted, replace the old tube and go to the next suspected stage.

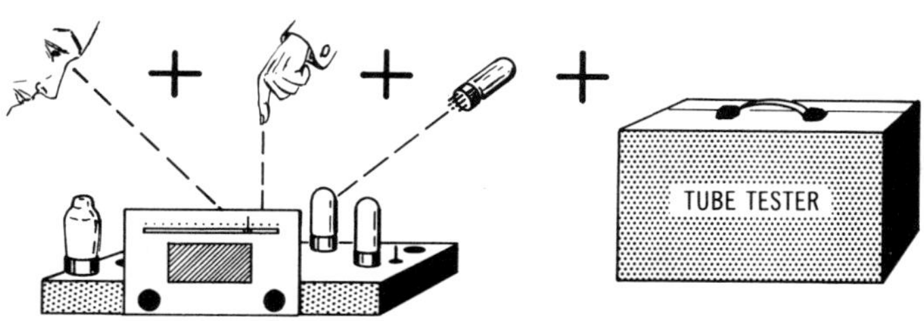

Fig. 5-7. Tube-testing methods.

Q5-5. Which of the above methods (in addition to a tube tester) can be used to determine whether the heater of a glass vacuum tube is operating?

Your Answer Should Be:

A5-5. All three—sight, touch, and substitution—can be used.

TUBE TESTERS

Visual, touch, and substitution tests are useful methods of identifying defective tubes. However, a test instrument could make these checks more rapidly. Such an instrument is called a **tube tester** or **tube checker.**

What Will a Tube Tester Check?

A tube tester can check almost all the characteristics of a tube that you may desire to know. Expensive laboratory models are designed that duplicate the operating conditions of the circuit in which the tube is to be used. In other words, the tube tester is capable of imitating the circuitry so closely that it is, in effect, a controlled substitution check.

Although expensive, this type of tube tester gives the kind of positive check that only actual circuit conditions make possible. A single tube type can be used in many different kinds of circuits. Within each of these circuit types, there are many variations in circuitry and applied voltages. An example of this is shown in Fig. 5-8.

The only true way to determine whether the same triode will operate in either one of these cathode-follower circuits is by actual trial in the circuit and observation of its performance. Some laboratory models of tube testers approach this capability, but the simpler types do not.

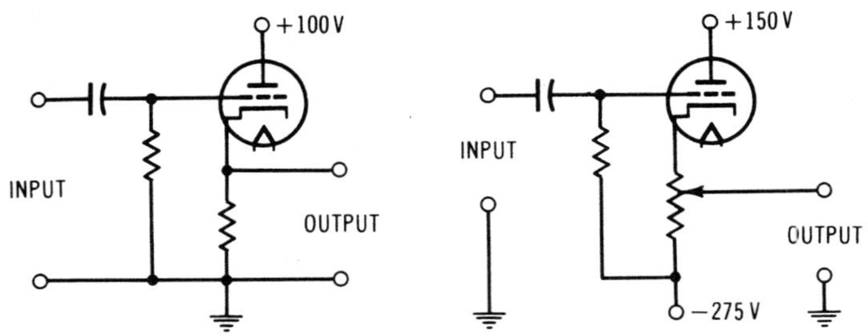

Fig. 5-8. Cathode-follower circuits.

Practical Tube Testers

There are two variables of practical tube testers—**emission** testers and **mutual-conductance** testers.

Emission Tester—As its name implies, the emission tester measures the ability of the cathode to emit electrons. Although a defect in cathode emission is one of the more frequent causes of tube failure, it does not tell the full story about a tube. In the first place, the tester can only give a fair approximation of the life left in the cathode. Secondly, an emission test will not reveal the ability of single and multigrid tubes to amplify.

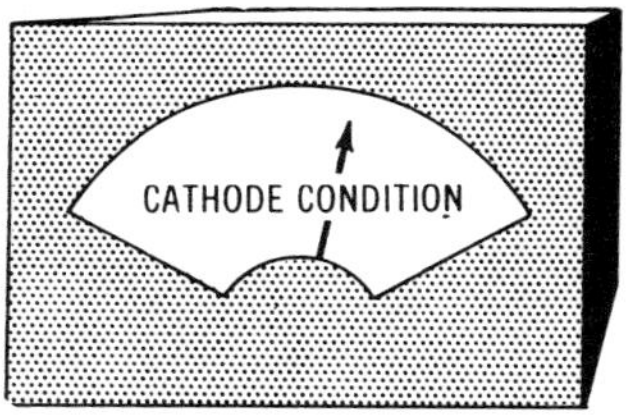

Fig. 5-9. Purpose of an emission tester.

Mutual-Conductance Tester—The terms mutual conductance and transconductance both have the same meaning. The term mutual conductance is usually used when discussing tube testers.

The mutual-conductance tester measures the amplifying ability of a tube having grids. Mutual conductance is an electron-tube rating. It is equal to the change in plate current divided by the change in grid voltage that causes the plate current to change (if the amounts of change are small and the plate voltage is constant).

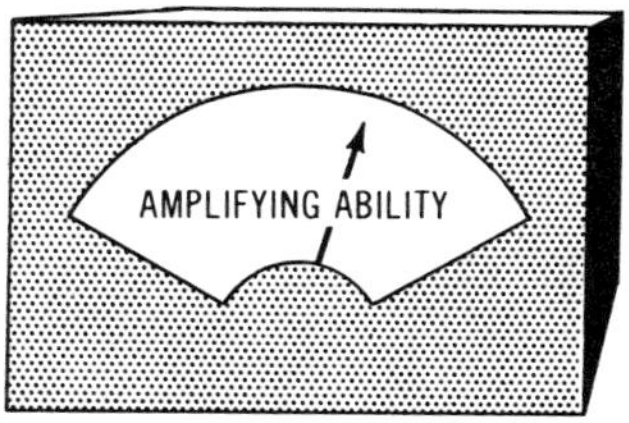

Fig. 5-10. Purpose of a mutual-conductance tester.

Q5-6. What is the only positive test for a vacuum tube?

Q5-7. Will an emission tester give a fair test of a diode?

Your Answers Should Be:

A5-6. The only positive check for a vacuum tube is to determine whether or not it will operate properly **in its designated circuit.**

A5-7. **Yes.** Since a diode does not amplify (has no grid elements), an emission test provides a fair indication of its operating quality.

Measuring Mutual Conductance

The mutual-conductance tube tester measures the amplification factor of a tube by solving the formula for mutual conductance, g_m.

$$g_m = \frac{\Delta I_p}{\Delta E_g} \text{ (with } E_p \text{ constant)}$$

where,

ΔI_p is a small change in plate current,
ΔE_g is the change in grid voltage that causes I_p,
E_p is the plate voltage.

Mutual conductance is the ratio of a small change in plate current to the small change in grid voltage that produced it. In effect, a mutual-conductance tester holds the plate voltage constant while changing the grid voltage. It then measures the change in plate current that takes place. The resulting measured value in **micromhos** can be compared against the average value for the tube type.

There are two types of mutual-conductance testers. One is called the **absolute,** or **direct-reading,** type. Its circuitry is designed to measure changes in plate current so closely that it can give a direct reading on a meter calibrated in micromhos. The other type, called **a relative,** or **dynamic,** tester, approximates the change taking place and provides a reading on a scale containing words or symbols.

The tube tester most frequently used by technicians is the dynamic mutual-conductance type. It is not as expensive as the direct-reading model, but it is many times more useful than the emission tester.

Either type of g_m tester also tests for shorts, noise, and gas. The better emission testers also make these tests.

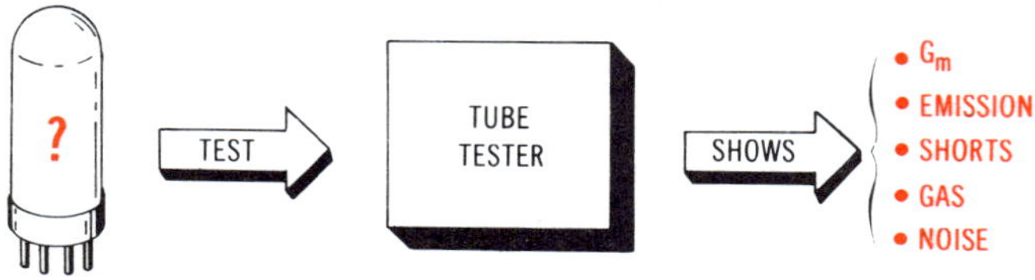

Fig. 5-11. Using a tube tester.

Essential Parts

Whether it is an emission or mutual-conductance type, a tube tester consists of a minimum of four main groups. These four areas are tube sockets, switches, test directions, and a meter. More complicated testers may contain additional functions.

Tube Sockets—Tube sockets are required to hold the tube in the tester. Since there are a variety of different tube bases, there should be one socket for each of the popular types. A mutual-conductance tester might have as few as eight or as many as fourteen sockets, depending on how many different tube bases it is designed to accept. An emission tester repeats the same tube sockets a number of times to decrease the number of required switches.

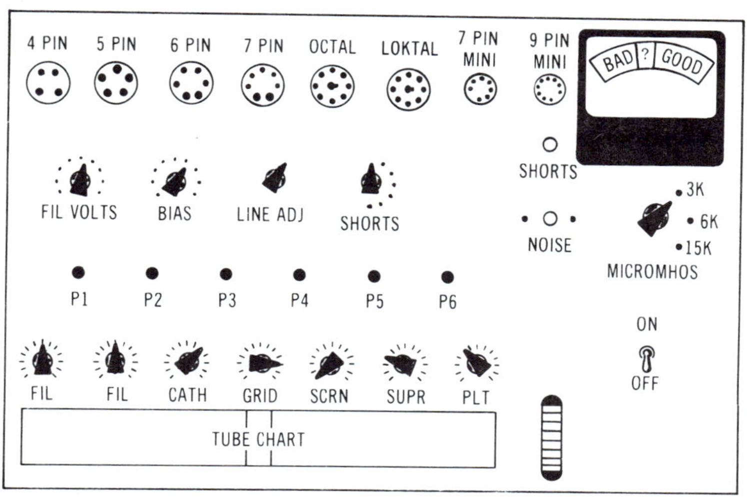

Fig. 5-12. A typical tube tester.

Q5-8. Will a dynamic tube tester test for emission?

Q5-9. What are the four basic areas in a tube tester?

Your Answers Should Be:

A5-8. **Yes.** Since it measures the transconductance of a tube, the dynamic tube tester also indicates cathode emission capability.

A5-9. **Tube sockets, switches, test directions,** and a **meter** are the four basic areas in a tube tester.

Switches—An emission tester can have as few as one or two switches. This tester is limited to making plate-current readings and, possibly, checking for shorts or gas. A mutual-conductance tester has several switches. Rotary wafer switches can be used to connect standard voltages to the appropriate terminals of the socket, depending on the pin-number arrangement (filament, grid, plate, etc.) of the tube being tested. Other switches are used to check filament volts, bias, line adjust (to zero meter), shorts (checking each pin with respect to the others), and micromhos. Switches also include several push buttons (to select the desired test).

Test Directions—Directions for setting the switches are included in a manual that accompanies the tester or on a paper roll built into the tester, as shown in the lower left-hand corner of Fig. 5-12. A section of a typical chart is shown in Table 5-1.

Table 5-1. Tube Chart

		Selector Switches											
Tube Type	Fil Volts	FIL	FIL	CAT	GRD	SCN	SUP	PLT	BIAS	SCALE	PRESS	Mut Cond	Notes
12A4	12.6	5	6	2	9	1	2	1	25	6K	P2	4900	
12A5	12.6	8	7	4	2	3	5	1	50	3K	P2	1130	

Meter—All tube testers have a meter. Some have a scale that is divided into three areas marked GOOD—WEAK—REPLACE, or with similar words. Such a scale can be found on either an emission or mutual-conductance tester. Another meter may have its scale calibrated in micromhos to measure transconductance directly. This type of meter would be found only on a mutual-conductance tester. Some mutual-conductance

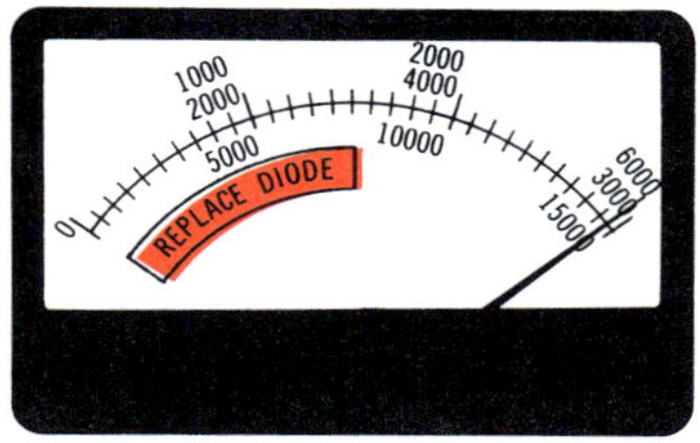

Fig. 5-13. Test meter.

testers have both micromho and word scales. Fig. 5-13 shows a typical meter.

HOW TUBE TESTERS OPERATE

The amount of plate current that flows in a tube can be an indication of the emitting capability of the cathode. The circuit in Fig. 5-14 provides this type of measurement.

Emission Test

In an emission test of grid-type tubes, the grid(s) and plate are connected together by switching or prewired tube sockets. The tube will then operate as a diode. Resistance R_1 is set at the proper value for each tube so that the meter will read GOOD, WEAK, or BAD. Since plate current will be pulsating dc, the meter will read its average value.

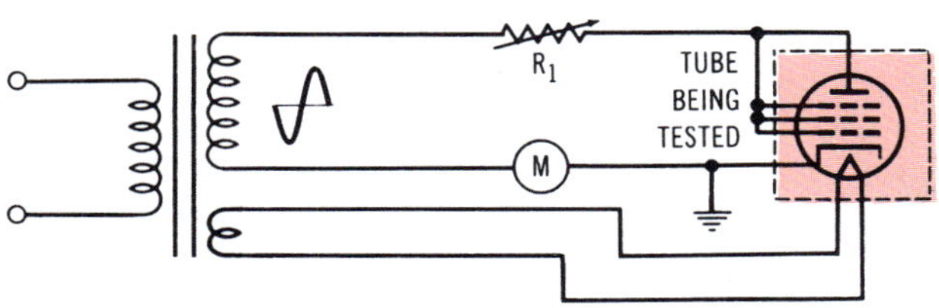

Fig. 5-14. Circuit for measuring plate current.

Q5-10. For a 12A4 tube, which switch would you set to 6K? Refer to Table 5-1.

Q5-11. What test would be made when P2 is pressed?

Q5-12. What is the purpose of the *Mut Cond* column on the chart?

Q5-13. What would the meter read if the grid and cathode were shorted?

Q5-14. What would it read if the cathode were open?

Your Answers Should Be:

A5-10. "6K" refers to a setting for the **MICROMHO switch.**

A5-11. Pressing P2 would connect the meter to read the **mutual conductance** of the tube.

A5-12. Each number under the Mut Cond column identifies the **average value in micromhos that a tube of a given type should indicate.** If the reading is below this number, the tube should be rejected.

A5-13. **Too GOOD.** Unless the meter or transformer were properly fused, the meter would be damaged by the excessive current flow. For this reason, **a tube should always be checked for shorts before being tested for emission or mutual conductance.**

A5-14. **BAD.** If the cathode were open, no current would flow.

Limitations of an Emission Test

The emission of a tube that indicates GOOD may be a false indication of how well the tube will operate. Sometimes a cathode will be highly electron-productive just before it goes bad. There may also be only one spot on the cathode that is highly emissive. Such a tube may give a good emission test but may not be good enough to operate in a circuit. Low emission does not necessarily indicate a bad tube. It may still have many hours of life left. A tube having grids may show good emission and yet not operate in a circuit.

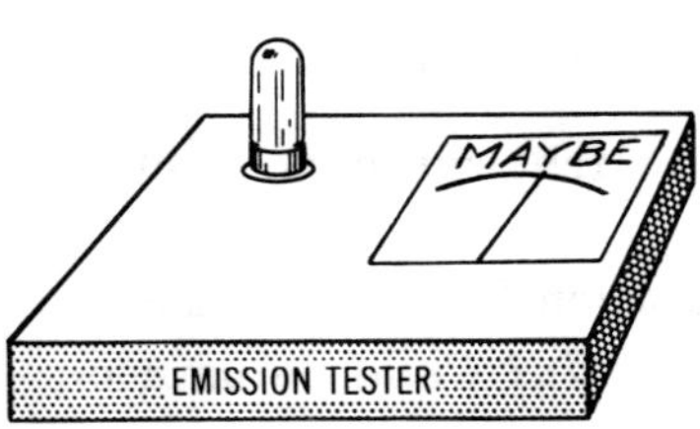

Fig. 5-15. Emission test has limitations.

Testing for Transconductance

As stated previously, mutual conductance has the same

meaning as transconductance. It is a figure of merit designed into a tube. Transconductance indicates the ability of a tube to amplify by specifying the change in plate current for a given change in grid voltage. A value of transconductance, expressed in micromhos, is assigned to each tube type in accordance with its design.

Transconductance will decrease in a tube as it is being used. Reduction of transconductance will result from the continued weakening of cathode emission and/or distortion of grid structure through periodic heating and cooling. A minimum value of transconductance can be assigned to each tube type. Any tube measuring below this value can then be considered to have fallen below the desired amplifying capability and should be discarded.

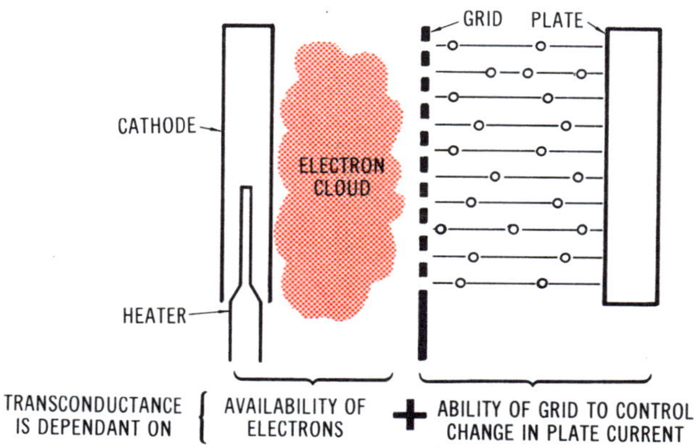

Fig. 5-16. Factors affecting transconductance.

Q5-15. What is transconductance?

Q5-16. Transconductance is expressed in terms of what units?

Q5-17. What factors reduce the transconductance of a tube?

Q5-18. Transconductance of a tube will _ _ _ _ _ _ _ _ _ as the tube is used.

Q5-19. Does a low emission reading always indicate a bad tube?

Your Answers Should Be:

A5-15. Transconductance is the **ratio of plate-current change to grid-voltage change.**

A5-16. Transconductance is expressed in terms of **micromhos.**

A5-17. The transconductance of a tube is reduced by a **weakening of the cathode emission and distortion of the grid structure.**

A5-18. Transconductance of a tube will **decrease** as the tube is used.

A5-19. **No.**

Circuit Used To Measure Transconductance

The tube being tested is V_1. Tube V_2 is a full-wave rectifier located in the tube tester. Its plates are each connected to equal secondary windings of a transformer. R_1 and R_2 are equal re-

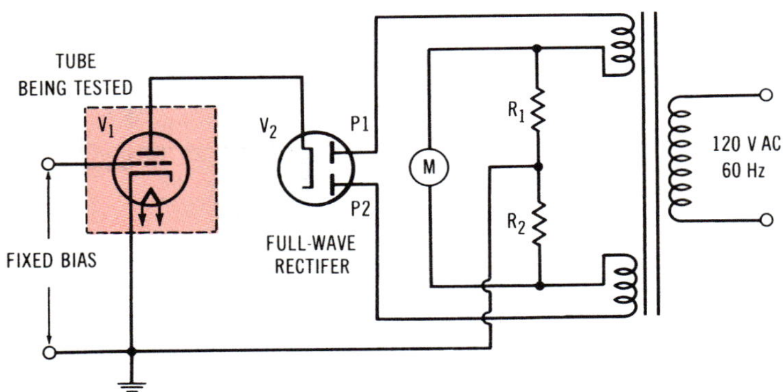

Fig. 5-17. Transconductance tester.

sistances shunted across the meter in the tester. When a fixed bias is placed on the grid of tube V_1, it will conduct through the rectifier. When voltage on the secondary is such that plate P_1 of the rectifier is positive and plate P_2 is negative, current will flow through resistor R_1 from top to bottom. The pointer in the meter will tend to swing in one direction. During the next half cycle, plate P_2 will conduct and current will flow through R_2 from the bottom to the top. The meter will tend to swing in the other direction. At 60 Hz, the net movement of the meter is zero.

An ac voltage can be applied to the grid of tube V_1 from a separate secondary winding. Suppose that this voltage makes the grid of V_1 less negative when plate P_1 is positive. During this half cycle, plate current will increase in V_1. A larger amount of current will flow through P_1 and resistor R_1, tending to swing the meter pointer a greater distance across the scale than before. During the other half cycle, plate current decreases in tube V_1. Less current flows through P_2 and R_2 than before, causing the pointer to tend to swing a smaller distance. Since the meter is passing more current in one direction than the other, the pointer will have a net deflection in one direction. The meter scale can then be calibrated in terms of the change of plate current that takes place with respect to a change in grid voltage.

Limitations of a Mutual-Conductance Tester

There are some limitations in the mutual-conductance tester. Manufacturers usually design their testers to show a tube defective when its transconductance has decreased to 70% of rated value. This figure is an average value. It is not valid under all conditions. Voltages applied to tube elements by the tester are only approximations of actual voltages applied in a circuit. Therefore, this is not a positive indication of how well the tube will operate in its designated circuit.

Testing for Gas

As was mentioned before, a small amount of gas is sometimes present inside the tube envelope. When electrons from the cathode strike the gas atoms, electrons from these atoms are separated from their nuclei. The atoms become positive ions. They are attracted to the negative grid and draw electrons from the grid circuit. Tube conditions are now no longer normal.

Q5-20. Can a transconductance test be made on a diode?

Q5-21. A tube is defective when its transconductance has decreased to approximately _ of its rated value.

Q5-22. What must a tube tester be able to measure to determine the transconductance of a tube?

Your Answers Should Be:

A5-20. **No.** A diode has no control grid, therefore, the meter would read zero.

A5-21. A tube is defective when its transconductance has decreased to approximately **70%** of its rated value.

A5-22. It must be able to measure **a change in plate current when a change in grid voltage is applied.**

A Circuit To Indicate the Presence of Gas

The circuit in Fig. 5-18, which is found in most tube testers, will determine the presence of gas. With the switch in position 1, the meter will indicate a value of plate current. When the switch is changed to position 2, there will be no change in the

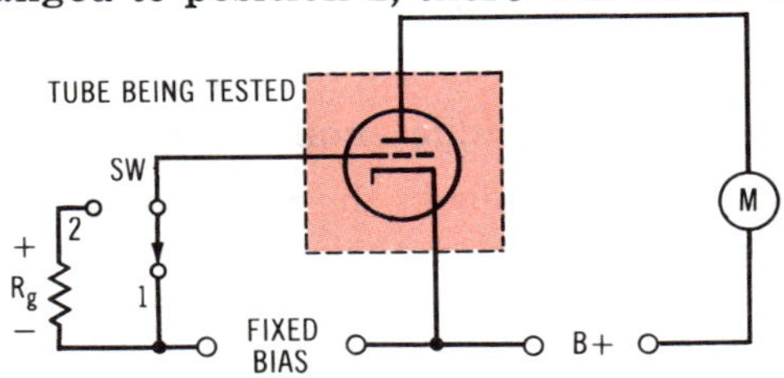

Fig. 5-18. Gaseous tube test circuit.

meter reading if very little or no gas is present. However, if positive gas ions are present, current will flow through R_g and develop a positive polarity from the grid to cathode. Bias will be reduced, plate current will increase, and an increased reading of the meter will be noted.

Testing for Noise

Noise is caused by loose electrodes, nonuniform electron emission, and heater-cathode current leakage. Nonuniform emission may or may not be detected by an emission or transconductance check. Tube substitution may be the only reliable check. If the equipment has an audible or visual output, tapping the suspected tube may verify the suspicion of loose electrodes. Some tube testers provide a pair of earphone jacks to permit such an audible check.

The tube in Fig. 5-19 is connected (switch position 1) as a diode, with the plate and cathode connected across the primary of the tester transformer. Plate current flows through the meter and a current-limiting resistance R_1. In position 2, the cathode is "floating." If there is no current leakage between cathode

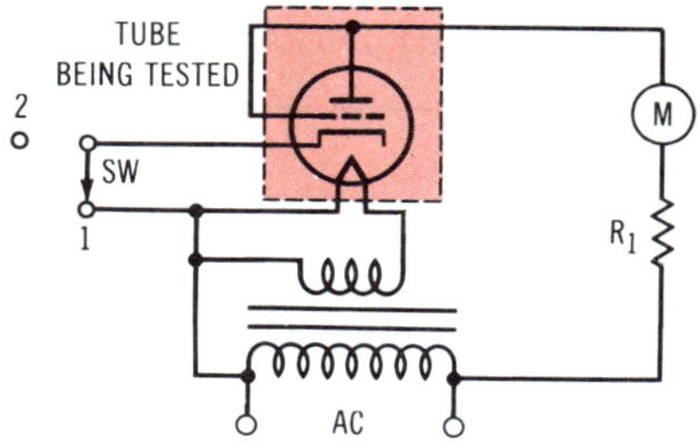

Fig. 5-19. Leakage test circuit.

and heater, the meter pointer will drop to zero. If there is leakage, the meter pointer will drop, but not to zero.

Testing for Shorts

When the cathode switch (Fig. 5-20) is in position 2, the cathode is connected to one side of the secondary through a neon bulb. All other switches are in position 1, connecting the tube elements to the other side of the secondary. Current will

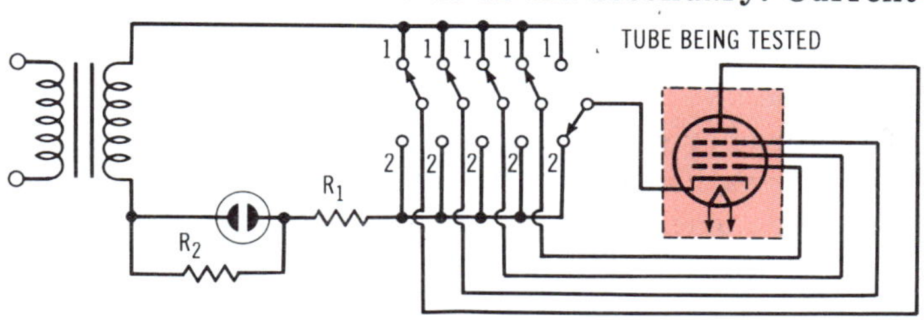

Fig. 5-20. Circuit used to test for shorts.

flow only if the cathode is shorted to any of the other elements. If current does flow, both halves of the neon bulb will light. If the cathode is not shorted, only half of the bulb will light during the half cycle when the cathode is negative with respect to the other elements. All elements can be checked one at a time in this manner. R_1 is a current-limiting resistance to protect the lamp while R_2 bypasses any small, stray, alternating currents around the lamp to prevent it from fully lighting when there are no shorted elements. Since tube elements may be loose, sharply tap the tube while making this check.

Q5-23. What factors cause a tube to produce noise?

Your Answer Should Be:

A5-23. The factors that cause a tube to produce noise are: **loose electrodes, nonuniform emission,** and **current leakage between cathode and heater.**

HOW TO USE A TUBE TESTER

A tube tester is not difficult to use. Follow the instructions contained in its manual, and apply them with common sense and a knowledge of how the instrument operates. Carefully follow instructions that apply to setting up the tester. In addition to the serious error of a false reading, improper settings could damage a tube and/or tester.

After proper settings have been made, insert the tube in the proper socket, turn the power on, and move the meter pointer to the setting prescribed with the **LINE ADJUSTMENT** control. Tubes should be allowed to warm up for at least one minute before making a test. Just before testing, make sure that the pointer is at its designated mark.

Make tests in the following order:

1. Test for shorts.
2. Test for cathode-to-heater leakage.
3. Test for noise.
4. Test for mutual conductance or emission.

WHAT YOU HAVE LEARNED ABOUT TUBE TESTERS

1. There are two general types of tube testers. An emission tester checks the capability of a cathode to produce electrons for plate current. A mutual-conductance tester approximates the transconductance (g_m) of a tube.
2. A GOOD reading on an emission tester can be invalid, since a cathode is capable of emitting large quantities of electrons just before it fails. A BAD reading can be false, since a cathode may emit electrons at a decreased but steady rate for a long period of time. Although an emission test may be suitable for diodes, it does not measure the true quality of grid-type tubes.

3. In a mutual-conductance tester, tubes are checked under voltage and circuit conditions that approximate but do not duplicate the exact operating conditions of the circuit in which the tube will be used.
4. A tube tester is made up of sockets (to test a variety of tubes), switches (to establish appropriate testing conditions for a particular tube), a meter (to provide a measurement of tube condition), and a chart or table (to provide switch-setting and testing instructions).
5. Meter scales on tube testers are of two general varieties. The better mutual-conductance tester has a scale calibrated in micromhos. Other mutual-conductance and most emission testers have scales containing words that are similar to GOOD—WEAK—BAD.
6. Most good tube testers also check for noise, gas, and shorts.
7. The only positive check of the quality of a tube is whether or not it will work properly in a specific circuit. This test is called a tube substitution check and can be made without a tube tester. Other checks that do not require a tube tester include visual and touch tests.

Q5-24. Name three types of tests that can be made without a tube tester.

Q5-25. Why is the mutual-conductance test a better indication of the condition of a triode than the emission test?

Q5-26. How could you determine, without using a tube tester, whether the heaters of a metal tube were working?

Q5-27. What tube test should you make first?

Q5-28. What does the mutual-conductance tester check?

Q5-29. What does the emission tester check?

Q5-30. A 12AU7 twin triode is connected in the amplifier circuit of a radio. The plate voltage (300 V) is kept constant. The transconductance of the tube is equal to 3100 μmhos. What is the change in plate current when the voltage applied to the grid changes from 0 V to −8.5 V? (Hint: $g_m = \Delta I_p / \Delta E_g$ when E_p is kept constant.)

Your Answers Should Be:

A5-24. The tests include **visual** (by eye), **touch** (by finger), and **substitution** (with a known good tube).

A5-25. Since mutual conductance is a measure of the change in plate current that will occur as the result of a change in grid voltage, **this test comes fairly close to measuring a tube under normal operating conditions.** An emission test will only measure the ability of the tube to pass plate current.

A5-26. **Metal tubes would feel warm,** sometimes very hot, if the heaters were working and the equipment were operating. Another check might be the use of an ohmmeter to measure for an open across the heater pins (with the tube not in the socket).

A5-27. Test for **shorts** first.

A5-28. A mutual-conductance tester measures **a change in plate current when a change in grid voltage is applied.**

A5-29. An emission tester measures the **ability of the cathode to emit electrons.**

A5-30. $g_m = \frac{\Delta I_p}{\Delta E_g}$

$\Delta I_p = g_m \times \Delta E_g = 3100\ \mu\text{mhos} \times 8.5\ \text{V} = \mathbf{26.4\ mA.}$

SEMICONDUCTOR TESTING

Semiconductors are relatively reliable devices. Some transistors, for example, are capable of operating for more than 30,000 hours. One example of this reliability is a digital computer that was tested during its development. The computer contained over 100,000 diodes and transistors. The test was run for two years, averaging 20 hours of operation per day. Within that period, there were only three semiconductor failures. Present-day computers use integrated circuits (ICs), and these show even greater reliability.

While vacuum tubes are the source of most troubles in the equipment in which they are used, semiconductors, particularly transistors, are relatively troublefree. However, a technician still must determine when a semiconductor device is operating properly.

Locating a Faulty Semiconductor

The approach to finding a bad semiconductor is the same as that for locating any other defective component. You do not test a component unless you have a good reason to suspect that it is defective.

Since most semiconductors are soldered into position, the preceding advice becomes even more meaningful. Soldering and unsoldering a number of transistors to find a suspected bad one can be a tedious time-consuming chore. Excessive heating can ruin a semiconductor. Therefore, be sure there is a good reason for removing a transistor before doing so.

Substitution Test—When you are reasonably sure you have found the circuit containing the trouble, and that the trouble is, in fact, a semiconductor, you can verify and correct the trouble by a substitution test. As with vacuum tubes, this is probably the simplest and most reliable of all tests. When the substitution of a good diode or transistor has restored the circuit to proper operation, you know that the semiconductor that it replaced was the cause of the trouble. Be very careful when removing and replacing semiconductors. Although strongly constructed, semiconductors are sensitive to excessive voltage, current, and heat.

When soldering or unsoldering a semiconductor, use the minimum heat required. Keep the semiconductor away from the chassis. Use a low-voltage soldering iron (30 to 40 watts) and a heat sink, as shown in the illustration of Fig. 5-21. A heat sink is a device for dissipating heat.

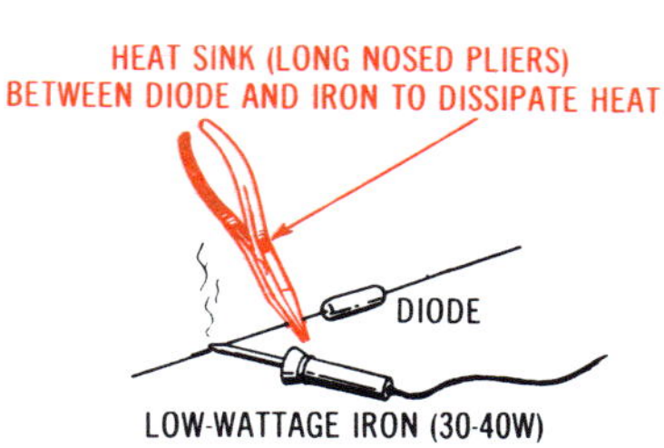

Fig. 5-21. Pliers used as a heat sink.

Q5-31. What factors can ruin a semiconductor?

> **Your Answer Should Be:**
>
> **A5-31.** Excessive **current, voltage,** or **heat** can ruin a semiconductor.

When a semiconductor is known to be defective, check the circuit for any defects that may have caused the damage. If the defects are not eliminated, they will also damage the substituted unit. These checks can be made with a voltmeter and an ohmmeter. Compare readings obtained with those contained in the equipment manual.

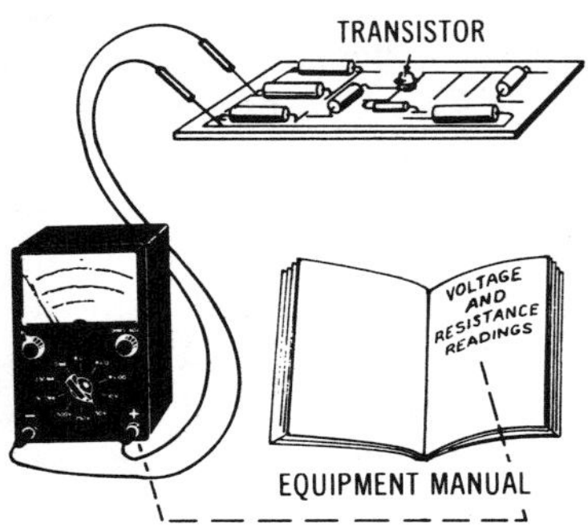

Fig. 5-22. Compare measured reading with equipment manual.

Crystal-Diode Tests

A crystal diode is a semiconductor. In the crystal-diode family are general-purpose germanium and silicon rectifiers (diodes) and silicon diodes constructed for high-power or very high-frequency purposes. Although these diodes may be effectively tested only under circuit-operating conditions, other tests can be made.

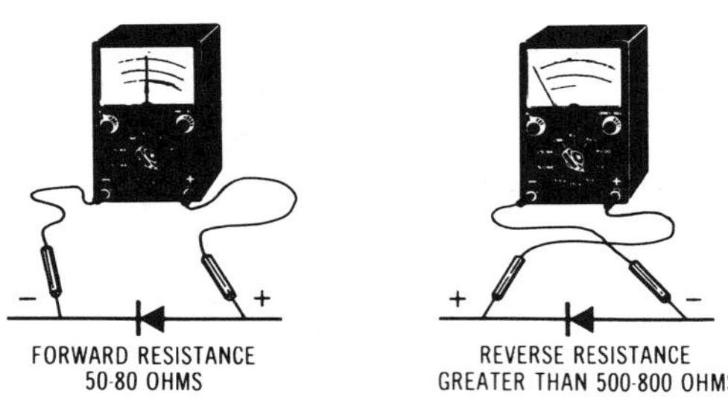

Fig. 5-23. Diode resistance measurement.

Resistance Measurement—A good diode will have a high resistance to current in one direction and a low resistance in the other. The ratio should be at least 10 to 1 for the diode to function as a rectifier. This is called a **reverse-to-forward** (sometimes **back-to-front**) resistance ratio, with the greater value being in the reverse direction.

Diode Test Set—Test sets are available to check the rectifying qualities of a diode. Some of them provide a combination resistance and current test. When the set is used as an ohmmeter, it will measure forward and reverse resistance. When it is used as a milliammeter, it will measure forward and reverse current. Other sets use one or the other method to obtain the forward-to-reverse ratio. Most sets are constructed as shown in the diagram of Fig. 5-24.

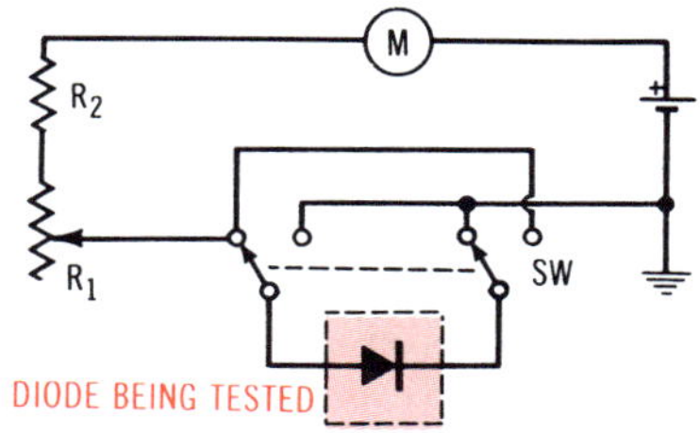

Fig. 5-24. Diode tester.

The diode is inserted in the device, and resistor R_1 is used to adjust the meter reading so that the pointer will remain on scale when the switch is thrown. The switch reverses the current direction through the diode. Resistor R_2 limits the current to a safe value.

Transistor Testing

There are laboratory and service instruments that measure transistor characteristics under both out-of-circuit and in-circuit conditions. Test sets of lesser capabilities are used daily by technicians who are concerned with repair, rather than design, of transistor equipment. However, many worthwhile checks can be made without the use of a transistor tester. When trouble occurs in transistor equipment, isolate the source of the trouble to a specific circuit before touching a single transistor, as diagrammed in Fig. 5-25.

Q5-32. The reverse-to-forward resistance ratio of a diode should be at least _ to _ .

Your Answer Should Be:

A5-32. The reverse-to-forward resistance ratio of a diode should be at least **10 to 1.**

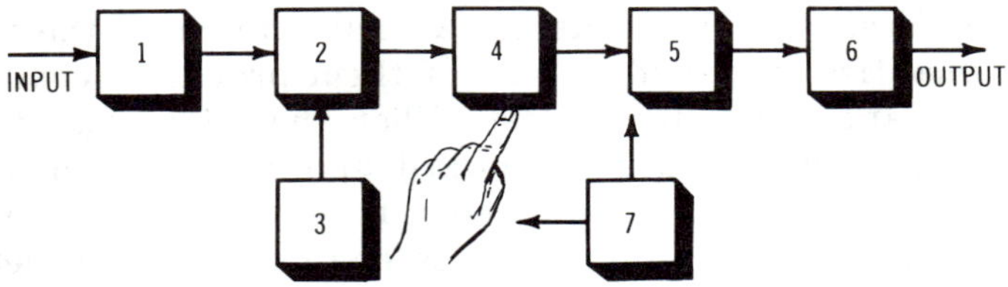

Fig. 5-25. Isolate the trouble before you try to fix it!

Transistor Testing Precautions

If the equipment manual contains waveforms at test points on a schematic or block diagram, an oscilloscope should be used to locate the circuit that is causing the trouble. If there are no waveforms available, voltage and resistance readings can be used to achieve the same results. As mentioned before, put off actual testing of a transistor until you are sure that it needs testing. When the faulty stage is isolated, use a multimeter to test the other parts of the circuit to determine whether abnormal conditions exist.

The multimeter used should have a high impedance to prevent loading the circuit. Low impedance across a circuit will change resistance and current values and provide a false voltage reading. The battery voltage of the ohmmeter used should not be in excess of 3 volts. Larger voltages may send an excessive amount of current through the transistor.

Parts and connecting leads are usually mounted very close together in the construction of transistor circuits. The metal tips of test probes are long enough to short-circuit leads when

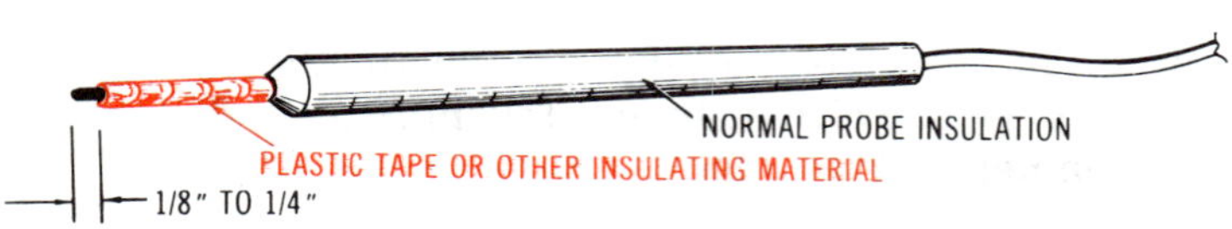

Fig. 5-26. Insulate the test probe.

taking measurements. If this happens, excessive circuit voltage or current could be shunted to another part. To prevent this, insulate the metal portion of the probes so that only a short portion of the tip is exposed.

When using a test instrument such as an ac voltmeter that could have a capacitor in series with the test lead, ground the probe to make sure the capacitor is discharged. If the capacitor were to discharge through a transistor, the transistor would likely be damaged.

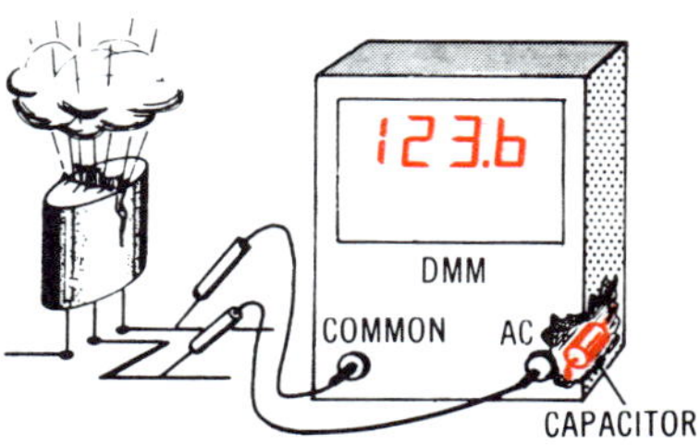

Fig. 5-27. Discharge capacitors in test equipment before using unit.

During the circuit-checking tests, it is best to remove the transistor if it is the plug-in type. However, removal is not absolutely necessary if you exercise necessary care.

Soldering irons, that are used to replace transistors and other semiconductors, should be of the low-wattage and isolated (or grounded) type, or they should be battery operated.

The technical manual for the equipment should supply sufficient information to make circuit checks that include the transistor. These tests involve the amount of bias or voltage applied to and the current through the elements of a transistor. With the exception of some circuits (pulse and power-amplifier stages, for example), transistors are usually biased so that ½ to 3 milliamperes of current flows through the emitter, and the voltage from collector to base is usually 3 to 15 volts.

Polarity, as well as the value of voltages, is particularly important to the safe operation of transistors. Check the amount of voltage, first, to be sure that it is not too high. Then, determine if the polarity is in the correct direction. Voltage on pnp transistors must be negative on the collector and positive on the emitter with respect to the base. Polarities in npn transistors are the reverse of those used in pnp types.

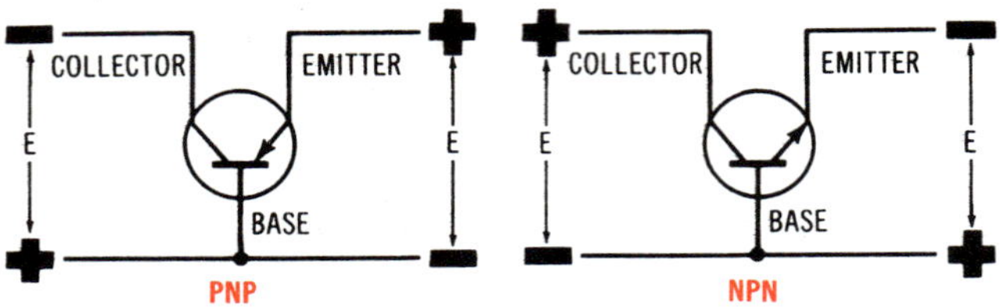

Fig. 5-28. Transistor voltage polarities.

Transistor Tests

If all checks indicate that the transistor may be defective and the transistor must be removed for further testing or replacement, turn off the power to the equipment. Removing or inserting a transistor in an operating circuit can cause the transistor to be damaged.

Ohmmeter Test—Forward- and reverse-resistance checks of the transistor elements can provide an indication of its condition. To make the test, ohmmeter readings (Fig. 5-29) should

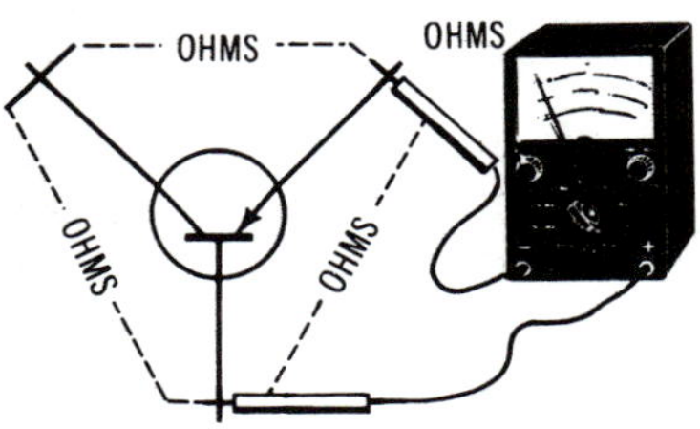

Fig. 5-29. Resistance test.

be taken in both directions (reverse the test leads) from emitter-to-base, collector-to-base, and collector-to-emitter. The purpose of the test is to determine whether any shorts or decreased resistances between elements have occurred. The large-to-small ratios for emitter-to-base and collector-to-base should be 500 to 1 or more. The direction of the ratio depends on whether the transistor is an npn- or pnp-type transistor. The resistance from collector-to-emitter should be nearly the same when measured in either direction. The two most common tests made by a technician are for leakage and gain.

Leakage Test—To determine leakage between elements, most testers place the transistor in series with a meter, battery, and current-limiting resistance. A good tester has a mi-

(A) Between collector and emitter. (B) Between collector and base.

Fig. 5-30. Measuring leakage in a transistor circuit.

croammeter. If current readings exceed those stated for the transistor, it should be discarded.

Gain Test—If the transistor leakage current is within suitable limits, gain can be tested. Gain is a measurement of the change that occurs in collector current as a result of a small change in base current. The reason for this test is that the current-gain capability of a transistor can decrease with age.

A variety of circuits are used to measure gain. All result in a ratio of I_c/I_b (collector current to base current). This ratio is matched against the minimum standard for the particular transistor. If the measured gain is too low, the transistor should be discarded. This test is often referred to as **direct-current gain,** since changes in dc current are involved.

Other Tests—More expensive sets measure the **punch-through** (or break-through) **voltage level** between collector and emitter to base. If current passes at a prescribed rise in voltage between the elements, the transistor would be considered defective.

An **ac-gain** test on a transistor is quite similar to a transconductance test for vacuum tubes. This test measures the ratio of collector-current change to emitter-current change as an **alpha amplification factor.** The ratio of collector-current change to base-current change is a **beta amplification factor.** Both ratios are measured with ac applied to the transistor and are then compared with desired values in a chart.

Integrated Circuit and Module Testing

There are no general-purpose IC testers because of the wide variety, different configurations, and nature of ICs. The same

Q5-33. What does a leakage test indicate?

Q5-34. What does a gain test indicate?

Your Answers Should Be:

A5-33. A leakage test reveals an **excessive flow of current between the elements** in a transistor.

A5-34. A gain test determines the **ratio of the change** in collector current to a corresponding change in base current.

applies to modules or circuit boards that contain a number of components arranged and designed for performing a particular function in equipment. The best test for these units is to check the input and output signals and the operating voltages. Normally, if a board is found to be defective, installing a replacement board is the usual practice.

WHAT YOU HAVE LEARNED

1. Semiconductors, particularly transistors, are reliable and have a long life expectancy.
2. A crystal diode is a semiconductor that allows current to pass more readily in one direction than the other.
3. Transistors are sensitive to excessive heat, voltage, and current.
4. Bias voltages and collector current can be measured in a transistor while it is still in a circuit. Measurement of reverse and forward resistances between the separate elements will indicate a defective transistor.
5. A diode tester is designed to accurately determine reverse-to-forward resistance ratios.
6. Transistor testers include circuits to measure collector leakage, dc gain, ac gain, and punch-through voltage.

6

Bridge Instruments

what you will learn

When you have completed this chapter, you will be able to explain the fundamental principles on which a bridge circuit is based. You will be able to determine how values of resistance, capacitance, and inductance can be accurately measured by a bridge instrument. On completion of this chapter, you will be able to use a bridge instrument after only a brief period of study of the individual bridge device that you intend to use.

WHAT IS A BRIDGE?

A **bridge** is a simple parallel circuit designed with a means of determining a potential difference between two parallel legs, as shown in the illustration of Fig. 6-1.

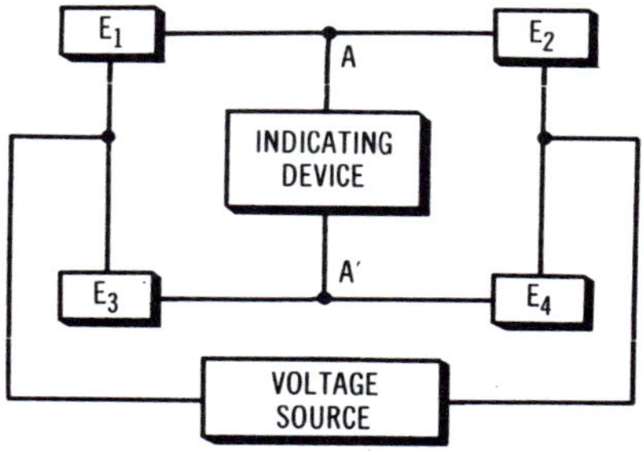

Fig. 6-1. Basic bridge circuit.

If voltage were applied across the bridge (parallel) circuit, current through the indicating device would be zero only when the voltages appearing at points A and A′ were equal.

Another fundamental principle is that the voltage across a parallel network is the same for either leg. In other words (disregarding voltage dropped in the conductors),

$$E_1 + E_2 = E_3 + E_4 = \text{Source voltage}$$

This means that a balance could be achieved between the two legs (no current flowing between A and A′) if the ratio of voltages on either side of A were equal to the ratio of the voltages on the respective sides of A′, or

$$\frac{E_1}{E_2} = \frac{E_3}{E_4}$$

HOW DOES A BRIDGE CIRCUIT WORK?

A brief review of circuit fundamentals may clarify this balance, or equality, between voltage ratios. In the parallel circuit shown in Fig. 6-2, $R_1 = R_2$. Disregarding the voltage drop in

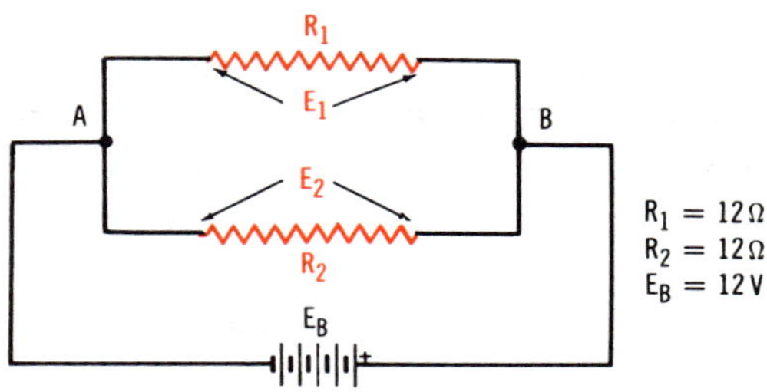

Fig. 6-2. Parallel circuit with equal resistances.

the conductors, the voltage difference between points A and B will be that of the source voltage (12 V). The current through resistances R_1 and R_2 will be 1 ampere each, or a total circuit current of 2 amperes. Since the resistances and currents in both legs are equal, E_1 is equal to E_2. Now divide one resistance into four equal parts (Fig. 6-3).

If the resistance, which shows a voltage difference of 12 V from one end to the other, were divided into equal quarters, each quarter would represent 3 V. Now divide both resistances

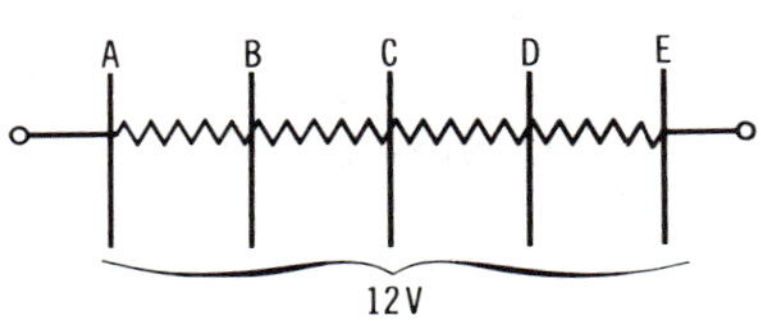

Fig. 6-3. Four equal parts.

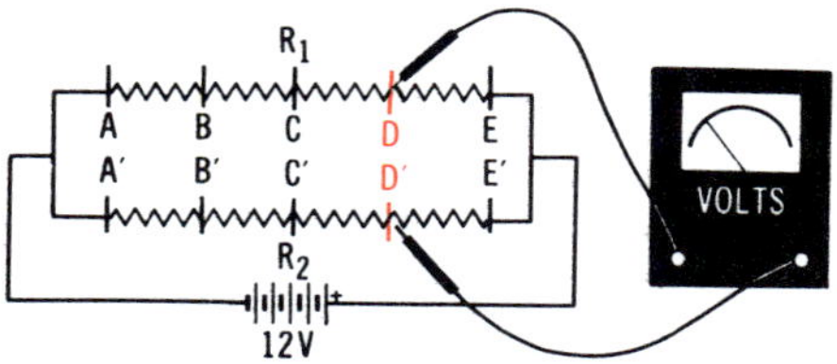

Fig. 6-4. Equal voltage points.

of the parallel circuit into equal quarters (Fig. 6-4). How much voltage would you measure between points D and D′? The answer is zero volts. The potential difference between points A and D is 9 V; this is equal to the difference between points A′ and D′. Therefore, no potential difference exists.

To show that the measurement works equally well from a reference point at either end, start from the other end of the network. The voltage at point B with respect to point E is −9 V; the voltage at D′ with respect to E′ is −3 V. Therefore, the difference of 6 volts will be measured from point B to D′.

Suppose that a parallel circuit had unequal resistances, as shown in Fig. 6-5. The voltage across resistances R_1 or R_2 would be identical, that is, 12 V. Again, assume quarter-section divisions.

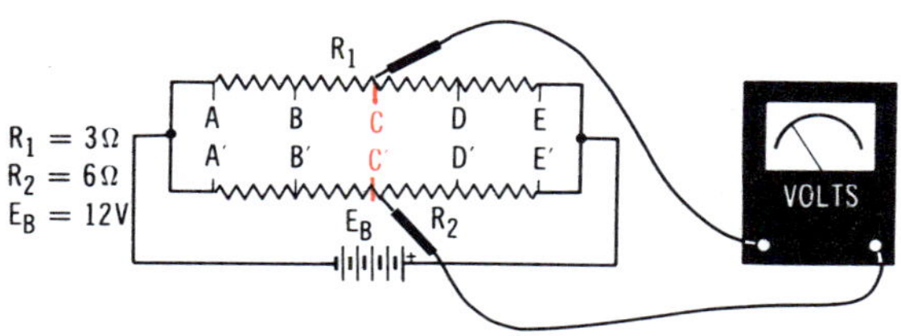

Fig. 6-5. Parallel circuit with unequal resistances.

Q6-1. What is the current through resistor R_1? Through R_2?

Q6-2. What is the value of resistance in each quarter section of R_1? Of resistor R_2?

Q6-3. What is the voltage difference between points A and B of resistor R_1? Between points A′ and B′ of resistor R_2?

Your Answers Should Be:

A6-1. The current through resistor R_1 is **4 amperes;** through resistor R_2, **2 amperes.**

A6-2. The value of resistance in each quarter section of resistor R_1 is **0.75 Ω**; of resistor R_2, **1.5 Ω.**

A6-3. The voltage difference between points A and B of resistor R_1 is **3 V**; between points A′ and B′ of resistor R_2 is **3 V.**

$$I_1 \times R_1 = 4 \text{ amperes} \times 0.75\ \Omega = \mathbf{3\ V}$$
$$I_2 \times R_2 = 2 \text{ amperes} \times 1.5\ \Omega = \mathbf{3\ V}$$

Voltage Relationships in Parallel Circuits

For every point on one parallel resistance leg, there is a point on the other leg that is at the same potential. This is true whether or not the total leg resistances are equal.

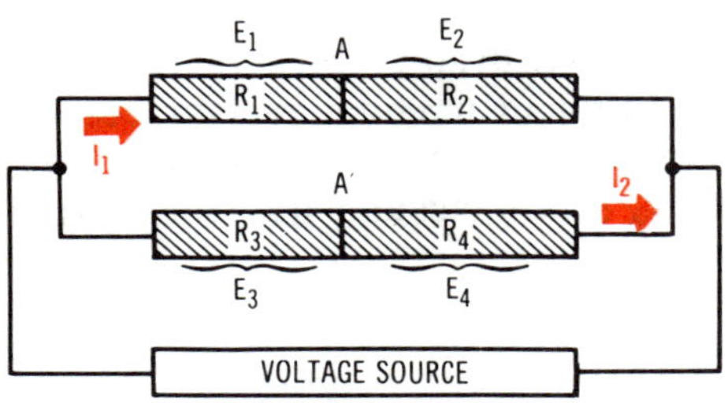

Fig. 6-6. Voltage relationships.

The voltage across a parallel network is the same for either leg. You can state that $E_1 = E_3$ and $E_2 = E_4$, in Fig. 6-6, if A and A′ are at the same potential. These voltage relationships can be stated in terms of equal ratios:

$$\frac{E_1}{E_2} = \frac{E_3}{E_4}$$

Since E is equal to IR, the above ratios can be stated in another manner:

$$E_1 = I_1 \times R_1 \qquad E_3 = I_2 \times R_3$$
$$E_2 = I_1 \times R_2 \qquad E_4 = I_2 \times R_4$$

therefore,

$$\frac{I_1R_1}{I_1R_2} = \frac{I_2R_3}{I_2R_4}$$

How could the IR equation be restated in terms of resistance only? In each ratio, I is the same, and it can be cancelled. Therefore:

$$\frac{R_1}{R_2} = \frac{R_3}{R_4}$$

Since the voltage drop across each resistor equals the leg current times the resistance, the resistances must have the same ratio as the voltages. The ratio can also be reasoned from the example shown in Fig. 6-7.

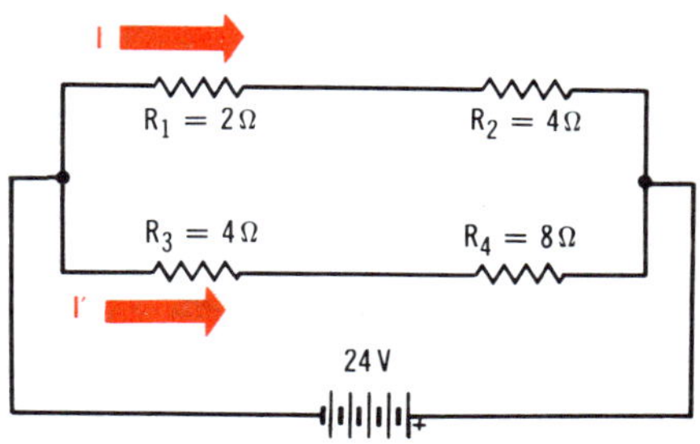

Fig. 6-7. Circuit for computing resistance ratios.

Both resistance ratios are equal to ½:

$$\frac{2}{4} = \frac{4}{8} = \frac{1}{2}$$

The current in each leg is:

$$I = \frac{24\ V}{R_1 + R_2} = \frac{24\ V}{6\ \Omega} = 4 \text{ amperes}$$

$$I' = \frac{24\ V}{R_3 + R_4} = \frac{24\ V}{12\ \Omega} = 2 \text{ amperes}$$

Both voltage ratios are ½ also.

$$\frac{IR_1}{IR_2} = \frac{I'R_3}{I'R_4}$$

$$\frac{4 \times 2}{4 \times 4} = \frac{2 \times 4}{2 \times 8} = \frac{1}{2}$$

Q6-4. Resistances in two parallel legs of a circuit will have a ____ potential difference between two proportional points on either leg resistance.

Q6-5. The voltage ratios in each leg of a parallel circuit must be equal to what?

Your Answers Should Be:

A6-4. Resistances in two parallel legs of a circuit will have a **zero** potential difference between two proportional points on either leg resistance.

A6-5. The voltage ratios in each leg of a parallel circuit must be equal to **the resistance ratios of the respective legs.**

RESISTANCE BRIDGES

If the ratios of two resistors in each leg of a parallel network are the same, the voltages at each of the corresponding junction points between resistance pairs are the same. Suppose the 8-ohm resistor in the illustration of Fig. 6-8 is changed to

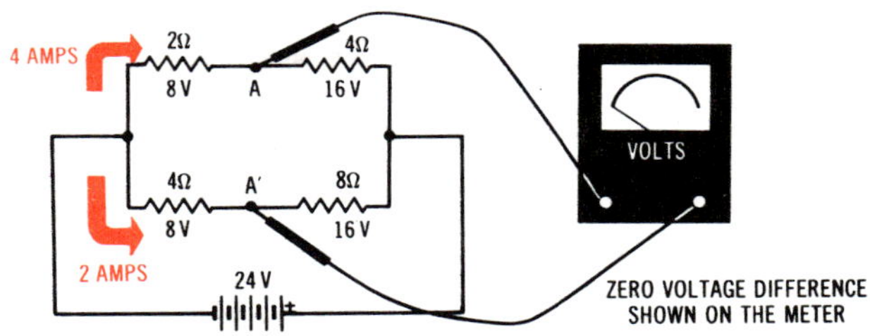

Fig. 6-8. Corresponding junction points.

4 ohms. The resistance ratios are no longer equal. The current in the lower leg is now 3 amperes, and the voltage at point A′ is 12 V, compared to 16 V at point A. The meter reads 4 V now. If you change the circuit to the one shown in Fig. 6-9, you have a means of measuring an unknown resistance.

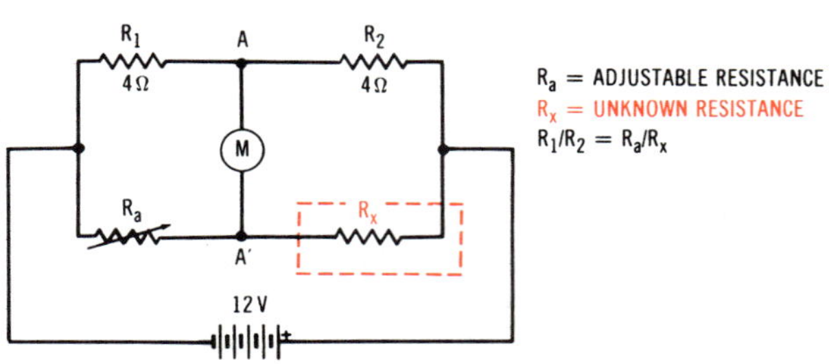

Fig. 6-9. Circuit for measuring an unknown resistance.

When the resistance ratios in the circuit legs are not equal, there will be a voltage difference between A and A′. Current will flow through the meter. When R_a is adjusted to register zero current, $R_1/R_2 = R_a/R_x$. If there were a means provided to show the value of R_a at this setting, how could you determine the value of R_x?

Assume R_a is equal to 13 ohms:

$$\frac{R_1}{R_2} = \frac{R_a}{R_x} \qquad \text{(Eq. 6-1)}$$

$$\frac{4}{4} = \frac{13}{R_x}$$

$$R_x = 13\ \Omega$$

If R_1 is 4 Ω, R_2 is 3 Ω, and R_a is 8 Ω, what is the value of R_x?

$$\frac{R_1}{R_2} = \frac{R_a}{R_x}$$

$$\frac{4}{3} = \frac{8}{R_x}$$

$$R_x = 6\ \Omega$$

Both problems were solved by selecting a value of R_x that would make the right-hand ratio proportional to the left. Since R_a was twice R_1, R_2 was multiplied by 2 to get R_x.

Multiply both sides of Equation 6-1 by both denominators. When ratios do not result in these convenient multiples, the solution becomes more difficult.

$$\frac{R_2}{1} \times \frac{R_x}{1} \times \frac{R_1}{R_2} = \frac{R_a}{R_x} \times \frac{R_2}{1} \times \frac{R_x}{1}$$

Cancel like terms in numerators and denominators on either side of the equality sign. This results in:

$$R_x \times R_1 = R_a \times R_2$$

Divide both sides by R_1 to obtain:

$$R_x = \frac{R_a \times R_2}{R_1} \qquad \text{(Eq. 6-2)}$$

Find R_x when:

Q6-6. R_1 is 200, R_2 is 400, R_a is 80.

Q6-7. R_1 is 80, R_2 is 20, R_a is 42.

Q6-8. R_1 is 30, R_2 is 70, R_a is 9.

Q6-9. R_1 is 40, R_2 is 56, R_a is 10.

Your Answers Should Be:

A6-6. 160 ohms. $R_x = \frac{80 \times 400}{200}$

A6-7. 10.5 ohms. $R_x = \frac{42 \times 20}{80}$

A6-8. 21 ohms. $R_x = \frac{9 \times 70}{30}$

A6-9. 14 ohms. $R_x = \frac{10 \times 56}{40}$

THE WHEATSTONE BRIDGE

You have, in effect, constructed a bridge for measuring an unknown resistance that is quite similar to an actual bridge. You have performed the mathematics required to find the value of the unknown resistance in the same manner as if you were actually using a bridge.

The most common type of bridge used for measurement is the Wheatstone bridge. Commercial models of this type of bridge can measure values of resistance from 1 ohm to 1 megohm with an accuracy of $\pm 1\%$. More expensive models can accurately measure resistances between 0.1 ohm and approximately 12 megohms.

Necessary components include a voltage source (usually a battery for resistance measurements), an indicating device (usually a sensitive galvanometer similar to those used in multimeters), accurate standard resistances that establish the ratio for measuring purposes (R_1 and R_2), a variable resistance for achieving a voltage balance between the two parallel legs (usually an accurate potentiometer with a scale on the front panel), and a means of connecting an unknown resistance into the bridge. All parts are designed for precise values to ensure the highest degree of accuracy. The meter is usually shielded to prevent stray fields from adding error or fluctuation to the reading.

Schematic Representation

In nearly all cases, the commercial bridge is shown schematically in a diamond shape. Both diamond and rectangular shapes are shown in Fig. 6-10.

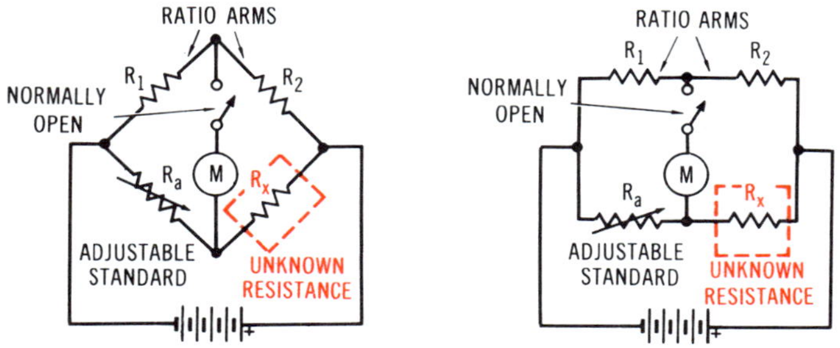

Fig. 6-10. Wheatstone bridges.

The meter is usually the type that has a center-scale zero. This permits the meter to record current going in either direction between the two parallel legs. A momentary-contact switch, in series with the meter, is normally open to keep the meter out of the circuit during the setting-up process. The switch is closed momentarily to observe meter deflection. The amount of deflection and its direction provide an estimate of how much resistance R_a should be changed. Adjusting R_a for a zero reading is often called adjusting for a **null.**

Bridge Operation

In the typical circuit of Fig. 6-11, resistance R_2 can be set at one of four positions to establish the R_1/R_2 ratio. In position 1, the ratio is 1/1; in position 10, it is 1/10, etc. Variable

Fig. 6-11. Resistance bridge.

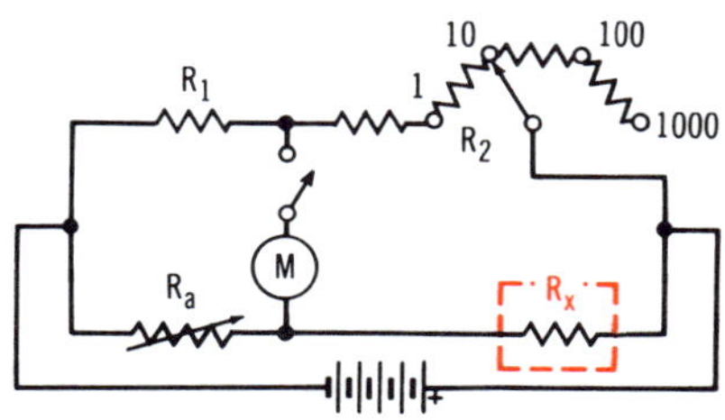

resistance R_a may be a single wirewound potentiometer with values calibrated on a front-panel scale, or it may be several **decade** (10-position) **switches** in series. One switch adds resistances in multiples of one, a second in multiples of 10, etc.

Q6-10. How accurate is a good Wheatstone bridge?

Q6-11. What is a decade switch?

Your Answers Should Be:

A6-10. A good Wheatstone bridge has an **accuracy of about 1%.**

A6-11. A decade switch has **ten calibrated contact positions.**

Operating the Wheatstone Bridge

In some bridges, the power supplies are variable types to provide low current for low resistances and high current for high resistances. In many models, the desired current is obtained by means of a current-limiting resistance.

Since the meter could be damaged if there were a large potential difference between the two legs, the shunt resistance connected to the meter is often made variable.

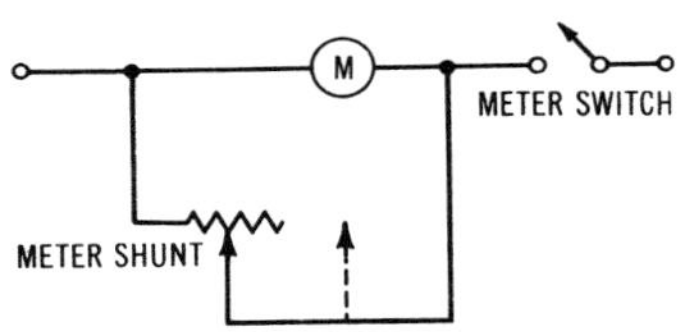

Fig. 6-12. Meter protection.

It is not often possible to make a reasonable guess as to the value of the unknown resistance. In such a case, the bridge ratio and resistance R_a cannot be set at values that are close to the unknown, thereby keeping the potential difference across the meter to a minimum. When securing resistor R_x to the measuring posts, the meter shunt should be moved to its lowest resistance. The meter switch is depressed, and meter deflection is noted. As resistance R_a and the bridge ratio are adjusted closer to the value of R_x (less difference of potential), the meter shunt is increased in resistance to permit greater sensitivity of measurement. During the final adjustment for a meter null (zero), the shunt arm is no longer contacting resistance, and all available current is flowing through the meter.

Another Resistance Bridge

There are variations of the basic Wheatstone bridge, such as the one shown in Fig. 6-13. In this bridge, the R_1/R_2 ratio is established by a potentiometer. The movable contact of the potentiometer is connected to a pointer on a direct-reading

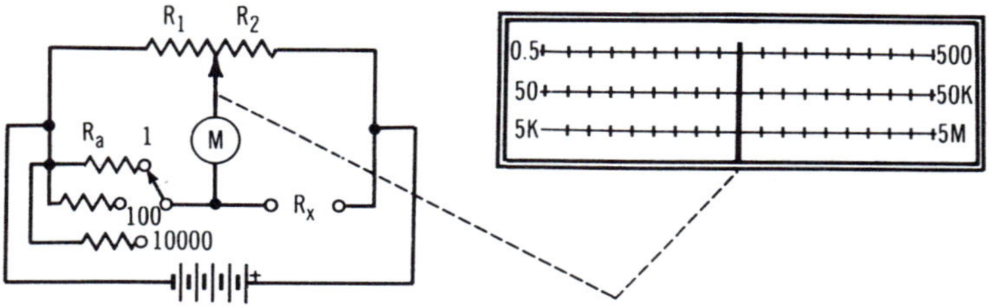

Fig. 6-13. A variation of the Wheatstone bridge.

scale. The potentiometer is adjusted for a null reading on the meter, and the value of R_x is read directly from the scale, in the range that is dependent on the setting of R_a. The three resistances of R_a increase in multiples to make it possible for a single pointer to serve all three scales. If R_a were 15 ohms in position 1 and 1500 ohms in position 2, it would be 150,000 ohms in position 3. Other multiples could be used, as well as a different number of resistances. The sample scale in Fig. 6-13 shows a bridge capacity from 0.5 ohm to 5 megohms, in three ranges.

Q6-12. Why is the power supply of a resistance bridge made variable?

Q6-13. How is this done?

Q6-14. What is meant by a meter null?

Q6-15. What method is normally employed to protect the meter from large voltages?

Q6-16. What is the value of R_x in Fig. 6-14A when R_a is set at 5 ohms?

Q6-17. What is R_x in Fig. 6-14B when R_a is 500 ohms?

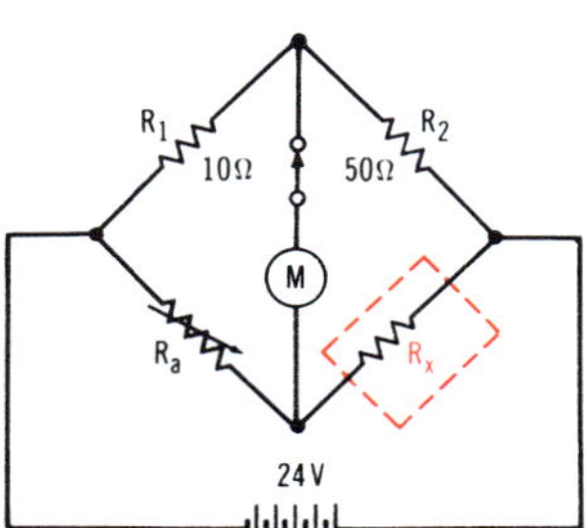

(A) For Problem Q6-16.

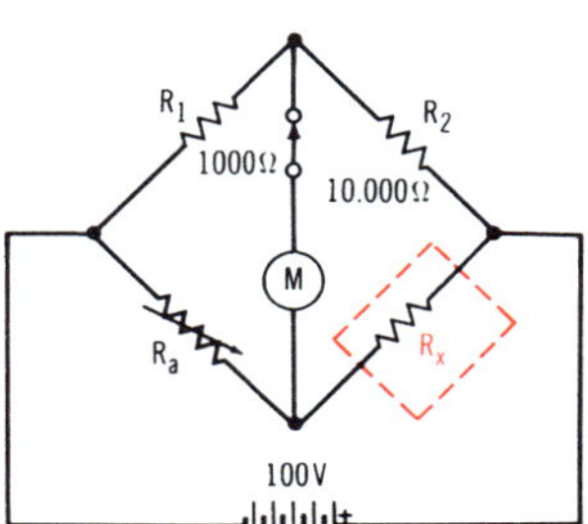

(B) For Problem Q6-17.

Fig. 6-14. Circuits for use with the problem questions.

Your Answers Should Be:

A6-12. The power supply is made variable **to provide the correct amount of current** for the various values of resistances.

A6-13. This is done by means of a **current-limiting resistance.**

A6-14. A meter null is that reading obtained when the **pointer indicates zero.**

A6-15. The **shunt resistance** connected to the meter is varied with respect to the amount of voltage applied.

A6-16. R_x equals **25 ohms.**

$$R_x = \frac{R_2 \times R_a}{R_1} = \frac{50 \times 5}{10} = 25 \text{ ohms}$$

A6-17. R_x equals **5000 ohms.**

$$R_x = \frac{10{,}000 \times 500}{1000} = 5000 \text{ ohms}$$

MEASURING CAPACITANCE WITH A BRIDGE

Using the same ratio principles, a bridge can also be made to measure the value of an unknown capacitance. The basic circuit is shown in Fig. 6-15.

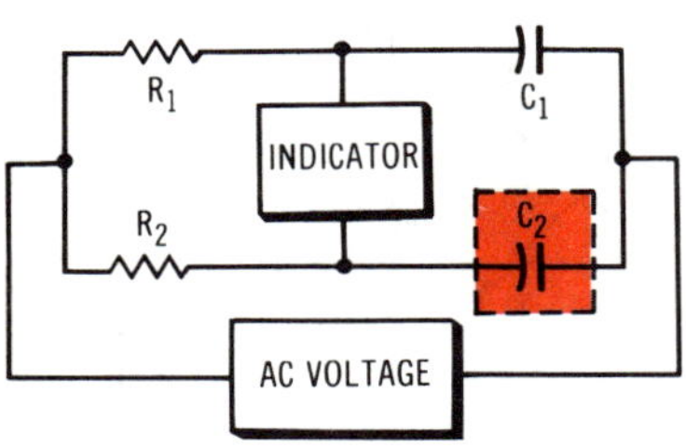

Fig. 6-15. Capacitance bridge.

A dc voltage source cannot be used because of the capacitors. As shown, the source must be an ac voltage. Most commercial models employ a frequency of either 60 or 1000 Hz. If the instrument is designed for a 60-Hz source, voltage may be taken directly from the power line. For 1000 hertz, some models have an oscillator built into the bridge, and others have provisions for connection to an external oscillator.

The indicator may be a built-in ac meter that allows current to pass through it in one direction only. Some units have jacks into which earphones can be plugged to permit listening for the null. Greater accuracy can be achieved by using an electronic analog meter, such as a solid-state vom or a vtvm.

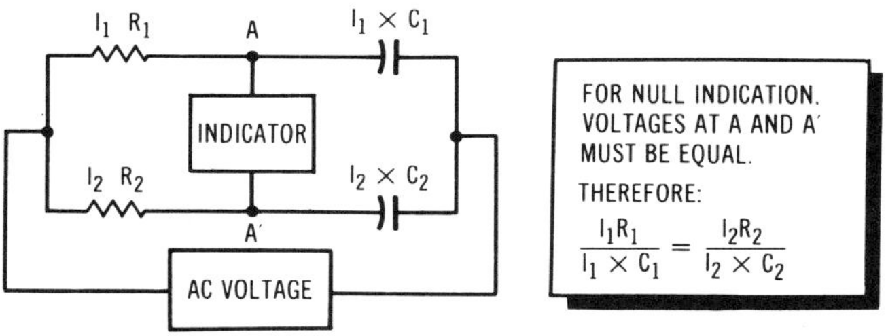

Fig. 6-16. Bridge ratios.

This is because the electronic analog meter can give a relatively large meter deflection even though the voltage difference between the legs of the bridge is small.

The principle of voltage ratios that you learned about a resistance bridge applies to the capacitance bridge as well. To obtain a null reading on the indicator, the voltage ratios in each leg of the parallel network must be proportional.

Q6-18. What type of voltage source must be used in the capacitance-bridge circuit?

Q6-19. What type of indicator offers the greatest accuracy in a capacitance bridge?

Q6-20. To obtain a null reading on the indicator, the voltage ratios in each leg of the bridge must be ______________ to each other.

Q6-21. If the voltage drop across one of the resistors is expressed as IR, how would you indicate the voltage across one of the capacitors?

Your Answers Should Be:

A6-18. An **ac voltage source** must be used in a capacitance bridge.

A6-19. An **electronic analog meter** is the most accurate indicator for a capacitance bridge.

A6-20. To obtain a null reading on the indicator, the voltage ratios in each leg of the bridge must be **proportional** to each other.

A6-21. **IX_C.** As you recall, voltage across a capacitor depends on its reactance, X_C.

Determining Capacitance-Bridge Ratios

As before, the current symbols cancel out, leaving ratios expressed in terms of resistance and reactance. If you wish to use the bridge for its intended purpose, reactance must be converted into capacitance. Capacitive reactance is:

$$X_C = \frac{1}{2\pi fC}$$

Before substituting the right element of the equation for X_C in the ratios and making the solution more complex than necessary, the ratios can be changed as follows:

$$\frac{R_1}{X_{C1}} = \frac{R_2}{X_{C2}} \quad \text{becomes} \quad \frac{R_1}{R_2} = \frac{X_{C1}}{X_{C2}}$$

Substituting for X_C in the right-hand ratio, you obtain:

$$\frac{1}{2\pi fC_1} \div \frac{1}{2\pi fC_2}$$

Inverting and multiplying:

$$\frac{1}{2\pi fC_1} \times \frac{2\pi fC_2}{1} = \frac{C_2}{C_1}$$

The $2\pi f$'s cancel out. The expression in terms of C is now:

$$\frac{R_1}{R_2} = \frac{C_2}{C_1}$$

The values of C are inversely proportional to the values of R in their legs of the circuit. This makes sense because a large capacitance has a small reactance.

Practical Capacitance Bridge

The schematic for a typical capacitance bridge is shown in Fig. 6-17. A close study of this illustration will reveal that it is essentially the same type of parallel circuit as those shown on the preceding pages.

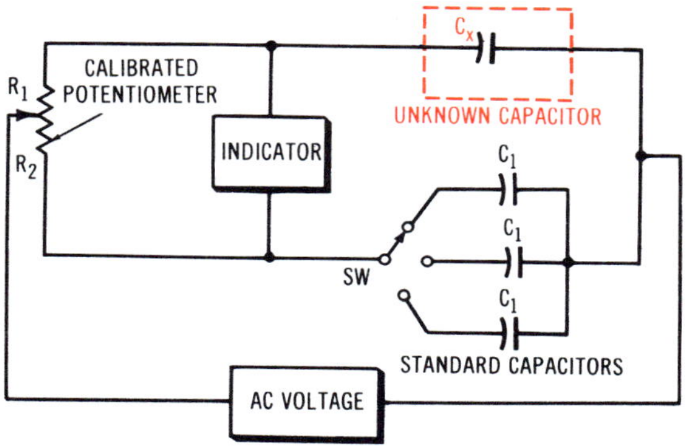

Fig. 6-17. Capacitance bridge.

By manipulating a switch, one of several capacitors can be selected as a standard, depending on the size of the unknown C_x. A calibrated potentiometer adjusts the bridge ratio (R_1/R_2) for a null reading on the indicator. With this arrangement, the arm of the potentiometer can be connected to a pointer that moves on a scale calibrated in capacitance values. This will be true only if all of the standard capacitors increase in value by the same multiple.

The bridge in Fig. 6-17 is suitable for measuring capacitors with little or no series resistance and no leakage between the plates. Significant values of either series or parallel resistance (leakage) will prevent balancing the bridge. This bridge can check most paper, ceramic, and mica capacitors.

Q6-22. What must the relationship of the standard capacitors be to permit adjustment of the potentiometer to proportional bridge ratios?

Q6-23. If the R_1/R_2 ratio is 1/10 and C_1 is 1200 pF, what is the value of C_2? (Refer to the circuit in Fig. 6-15.)

Q6-24. If the R_1/R_2 ratio is 1/100 and C_2 is 12 μF, what is the value of C_1?

Your Answers Should Be:

A6-22. The capacitors must increase in value **by the same multiple.**

A6-23. $C_2 = 120$ pF. $\frac{R_1}{R_2} = \frac{C_2}{C_1}$

A6-24. $C_1 = 1200\ \mu F$

Measuring Power Factor

Large capacitors have an internal resistance. This resistance must not only be compensated for in the bridge in order to obtain a null reading, but it also must be measured to determine the power factor of the capacitor. As shown in Fig. 6-18, the

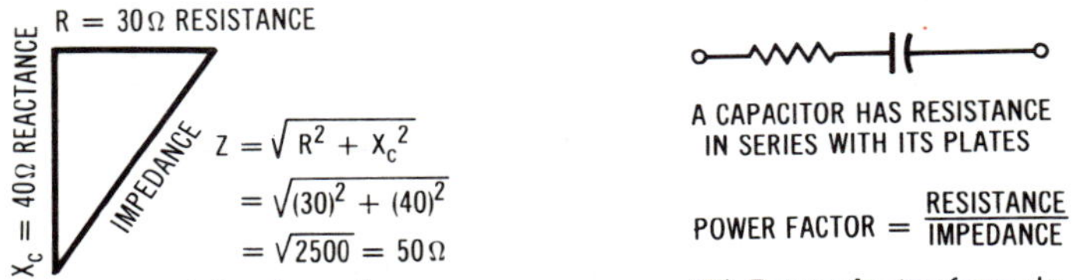

(A) Determining impedance. (B) Power factor formula.

Fig. 6-18. Power factor.

power factor of a capacitor varies in accordance with the size of its internal resistance. The diagram in Fig. 6-18A shows the relationship between the resistance and impedance elements of a capacitor; it also shows how impedance can be determined for the power-factor formula. If the internal resistance is zero, the power factor is zero. If the power factor is large (stated in percentage), it may seriously affect the operation of the circuit or device. Many bridges contain provisions for measuring the power factor. One method is shown in Fig. 6-19.

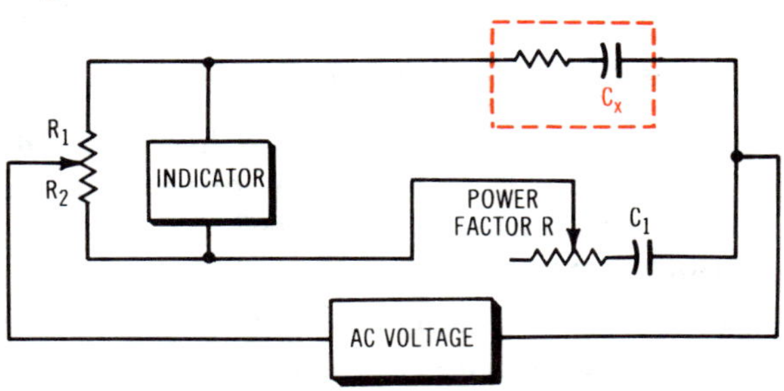

Fig 6-19. Bridge with provision for measuring power factor.

The R_1/R_2 ratio potentiometer is adjusted for a null indication as was done before, or as close as the resistance of C_x will permit. The variable resistance in series with C_1 is then adjusted for the final null reading. A front-panel extension of the variable resistance, probably labeled POWER FACTOR, will give the percentage of power factor.

Measuring for Capacitor Leakage

A leaky capacitor is one in which the dielectric, having lost some of its insulating quality, permits electrons to travel from

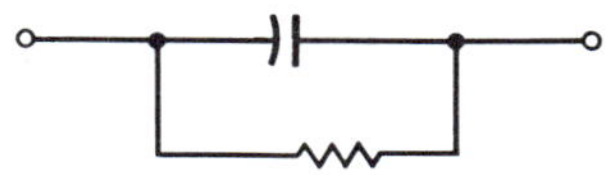

Fig. 6-20. Leakage.

one plate to the other. Leakage is considered to be a resistance in parallel with the capacitor.

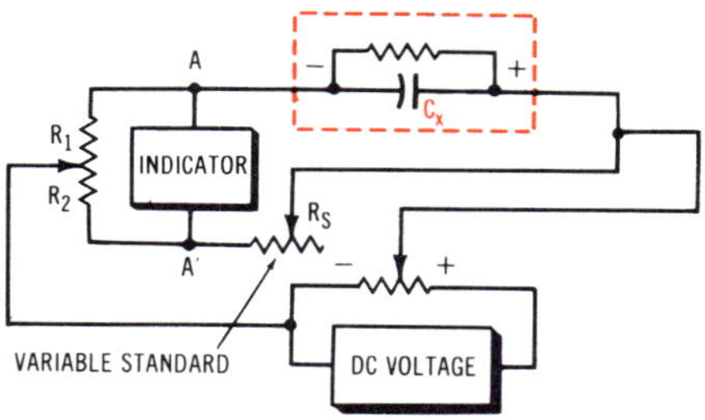

Fig. 6-21. Using a variable resistance and a variable dc voltage source.

Essentially, Fig. 6-21 is the same basic circuit as before. A switch on the front panel connects a variable resistance in place of C_1 and changes the voltage source to a variable dc. Voltage is set at the rated working voltage of the capacitor. An electrolytic capacitor will be damaged if polarities are not observed when connecting it to the measuring posts. Resistance R_s is adjusted to balance the bridge and obtain a null reading. If the measured resistance is low, leakage is high.

Q6-25. In the illustration of Fig. 6-21, the voltage at point A should be (greater than, the same as, less than) the voltage at point A′ to get a null reading on the indicator.

Q6-26. Why is a dc voltage used to measure capacitor leakage in the circuit of Fig. 6-21?

Your Answers Should Be:

A6-25. The voltage at point A should be **the same as** the voltage at point A′.

A6-26. If ac were used, shunting reactance would be added to the upper leg of the circuit. **The bridge could not be balanced using resistance R_s.**

MEASURING INDUCTANCE WITH A BRIDGE

Inductance can be measured with the bridge shown in Fig. 6-22. In previous bridges, a resistor was used for balancing a resistor, and a capacitor was used for balancing a capacitor.

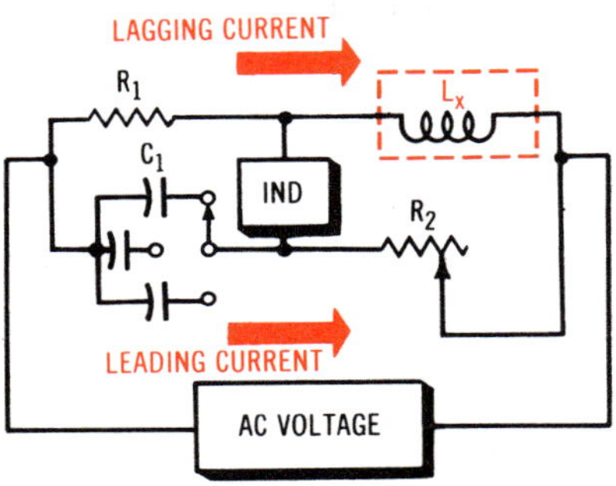

Fig. 6-22. Inductance bridge.

In this circuit, however, a capacitor is used to balance the inductance of a coil. A standard inductor could be used instead of a capacitor but this would cause a few undesirable conditions.

The illustration of Fig. 6-23 shows how the electromagnetic field around a coil can cut through a conductor and induce error-producing currents. The capacitor electrostatic field, however, is mostly contained within the capacitor. An addi-

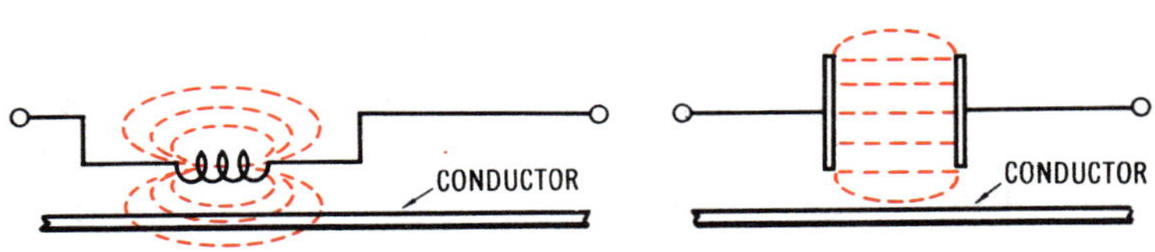

(A) Inductor field. (B) Capacitor field.

Fig. 6-23. Electromagnetic fields of a coil and a capacitor.

tional advantage of the capacitor is that it does not pick up stray fields as a coil does. If inductors were used to balance the bridge, they would be larger and more expensive than equivalent capacitors. The size and cost of the instrument would be greater. Finally, the same capacitors can be used for measuring both inductance and capacitance with the same bridge.

Using a capacitor opposite an inductor in a bridge, however, has one disadvantage. The current through the coil branch lags the current through the capacitor branch. The phase difference makes it impossible to balance the bridge. This condition can be compensated for by placing the unknown coil and the standard capacitor in opposite legs of the bridge across the indicator, as shown in the illustration of Fig. 6-22.

Typical Inductance Bridge

In the inductance bridge of Fig. 6-24 and the one just discussed, a null condition is established by the values of C_1 and R_2. With a proper value of C_1, R_2 (a variable resistance) is adjusted for a zero reading. The position of a pointer (on a front-panel scale) linked to R_2 indicates the value of L_x.

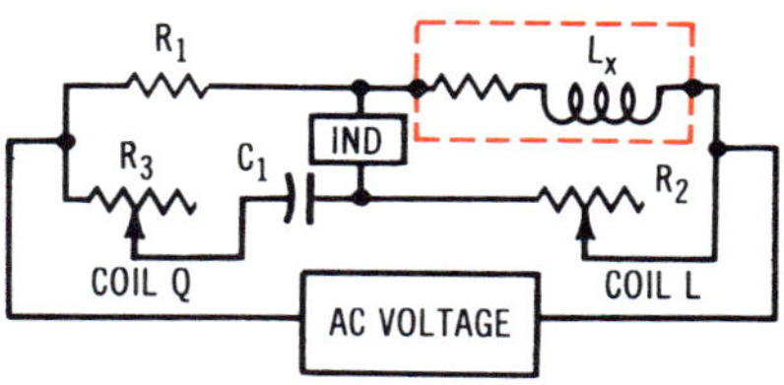

Fig. 6-24. Q and L measurement.

Measuring the Q of an Inductor

An inductor has a Q level, or figure of merit, that is its reactance-to-resistance ratio ($2\pi fL/R$). Resistance is that resistance of the wire used in the coil and it is considered to be in series with the reactance. Heavy wire and a small number of turns produce a high Q; a smaller wire and more turns produce a lower Q. Since Q, a measure of quality, is a factor used to determine the sharpness of resonance of a circuit which employs both L and C, it is desirable that the Q of the coil be measured. Resistance R_3 in the circuit of Fig. 6-24 is used for this purpose. Its scale can be calibrated in Q.

Q6-27. What is used to balance the coil inductance in an inductance bridge?

Your Answer Should Be:

A6-27. A **capacitor** is used to balance the inductance of the coil in an inductance bridge.

WHAT YOU HAVE LEARNED

1. A bridge is a test instrument capable of accurately measuring resistance, capacitance, inductance, and reactance. Although a bridge may not have a scale calibrated in reactance, you can determine reactance if you know C or L and the frequency.
2. The Wheatstone bridge utilizes the basic principle on which all bridge circuits are built. The arrangement of the bridge elements is such that the voltage ratios, between the elements in the parallel legs, can be adjusted to achieve a null (zero current) reading on an indicator placed between the parallel legs.
3. Each bridge requires a source of voltage, a parallel network containing sufficient variable elements to balance the bridge, an indicator to determine when the bridge is balanced (null indication), a scale to determine the ratio of the elements involved or a direct readout of the unknown value, and operating controls and jacks.
4. A resistance bridge contains a dc-voltage source, a sensitive dc meter, and a parallel network of resistances. When a condition of balance exists, the unknown resistance can be determined either by a direct reading from a calibrated scale or by calculation.
5. A capacitance bridge contains an ac-voltage source, a sensitive ac meter (or headphones), and a parallel network of resistance and capacitance.
6. With appropriate modifications, a capacitance bridge can measure the power factor and leakage of a capacitor.
7. An inductance bridge contains the same voltage source and indicating devices as a capacitance bridge, but its parallel network is normally made up of resistance, capacitance, and the unknown inductance.

7

The Signal Generator

what you will learn

There are occasions when a technician finds it necessary to apply a standard signal to an electronic circuit or device for testing or troubleshooting purposes. A signal generator is an instrument that serves as a source for such a signal. Within its desired range, it provides a signal having controllable frequency, amplitude, and modulation characteristics for testing or troubleshooting.

On completion of this chapter, you will be able to apply an understanding of basic electronic principles to the operation of a signal generator. You will learn how to effectively operate a signal generator after a brief study of its operating manual. You will also be able to use a signal generator in making troubleshooting checks.

WHAT IS A SIGNAL GENERATOR?

A signal generator is basically a transmitter. Although a transmitter may emit signals at a higher power and include

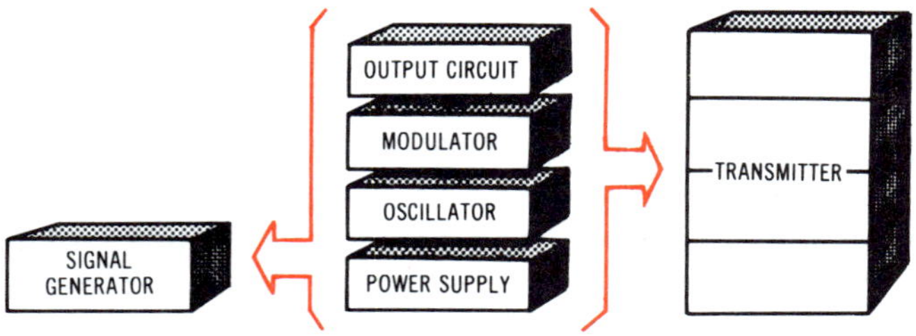

Fig. 7-1. A signal generator is a transmitter.

more circuitry than a signal generator, the basic functions of the two are identical. As shown in Fig. 7-1, the basic functions include a power supply, an oscillator, a modulator, and an output circuit. Each function may require a group of many circuits in a transmitter; in a signal generator, however, each function can be accomplished using only one or two circuits.

FUNCTIONAL UNITS IN A SIGNAL GENERATOR

Since the power supply of a signal generator is fairly standard, it will not be described. However, an understanding of how the other circuits work will be very helpful when using a signal generator.

Generator Oscillator

An oscillator is a circuit capable of generating a series of identical waveforms at some desired frequency (number of oscillations per second).

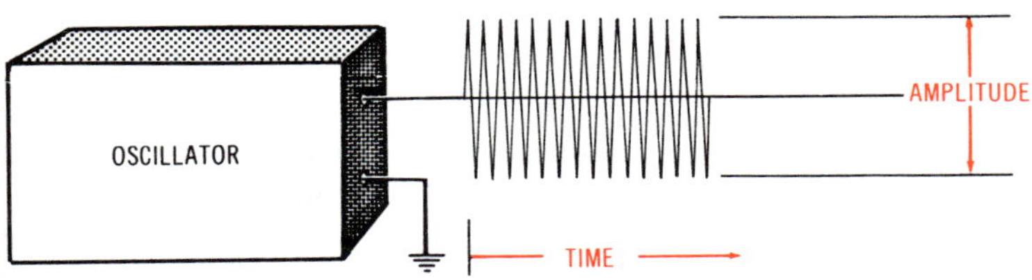

Fig. 7-2. An oscillator generates a series of identical waveforms.

Signal generators are classified in accordance with the frequency range of their oscillators. Audio-frequency (af) generators usually cover a range from about 20 to 20,000 Hz. Radio-frequency (rf) generators begin at 20,000 Hz and end at several thousand megahertz.

Modulation Oscillator

In a transmitter, the frequency generated by the oscillator is called a **carrier frequency.** Waves produced by a current of this frequency are capable of traveling many miles. The carrier wave has a constant frequency and amplitude; its purpose is to carry information, or intelligence, from a transmitter to a receiver. The carrier is altered by having another frequency, representing the intelligence, superimposed on it. For example, the standard broadcast band has carrier frequencies between 535 and 1605 kilohertz. Intelligence, in the form of sound (voice, music, etc.), is superimposed on the carrier and later taken from it at the receiving set.

The process of superimposing intelligence frequencies on a carrier frequency is called **modulation.** To be useful, an rf signal generator must be capable of modulating its carrier. It does so with an audio-frequency signal developed by a **modulation oscillator.**

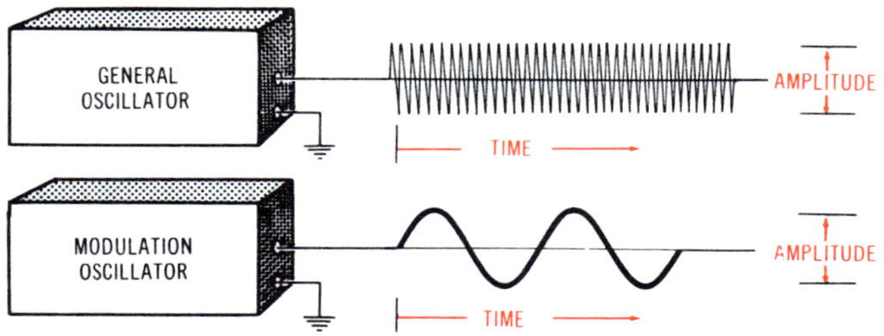

Fig. 7-3. A modulation oscillator.

Amplitude Modulation

There are several types of modulation. **Amplitude modulation** is the form used in standard broadcast transmissions (535 to 1605 kHz) and is defined as the process by which the carrier is varied in amplitude to resemble the amplitude and frequency of the intelligence.

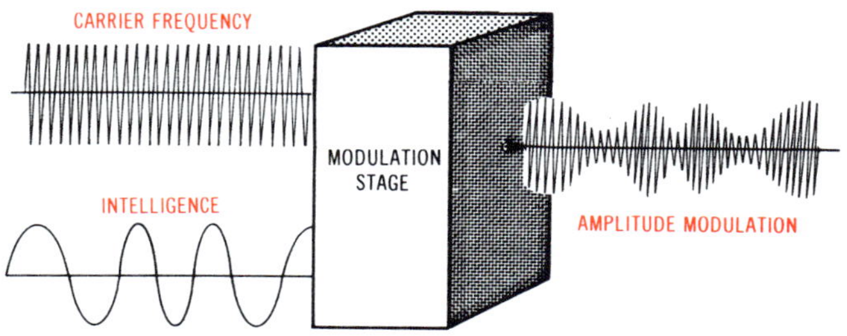

Fig. 7-4. Amplitude modulation.

Q7-1. What is the name given to the transmitted frequency that the generator oscillator duplicates?

Q7-2. Why is amplitude modulation the name given to the type of output waveform shown in Fig. 7-4?

Your Answers Should Be:

A7-1. Carrier frequency. It carries the frequency of the intelligence superimposed on it.

A7-2. The **amplitude** of the constant carrier frequency is **modulated** (varied) to conform with the frequency and amplitude of the intelligence signal.

Modulation Requirements

Although the rf oscillator must be capable of reproducing the desired carrier frequency, the modulation oscillator need not cover the entire audio range. In some signal generators, only a single modulation frequency is developed. Other signal generators may have two or three modulation frequencies or a variable modulation frequency.

Percent of Modulation—The amount that the carrier amplitude is varied is called the **percent of modulation.** It is controlled by the amplitude of the modulating signal.

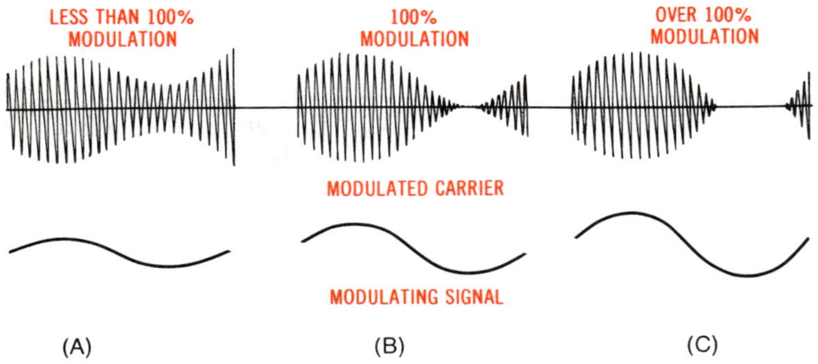

Fig. 7-5. Modulation percentage.

The amplitude of the modulating signal in Fig. 7-5B is such that the **envelope** (outline of modulation on the carrier) is caused to become zero for an instant before rising again. This is 100% modulation. Modulation is less than 100% in Fig. 7-5A. The exact percentage is determined by the relationship between the amplitudes of the carrier and modulating waves. Fig. 7-5C shows modulation greater than 100%. This causes distortion in reception.

Since there is little need for modulating an audio frequency, this feature is seldom found in af signal generators. Better models of rf generators always incorporate some form of modulating capability. In some rf generators, the amplitude of the modulation-oscillator signal can be varied; in others, it cannot. Some models permit an external audio oscillator to be used for modulation.

Frequency Modulation—Another method of superimposing an intelligence signal on the carrier is called **frequency modulation (fm).** In frequency modulation, the **frequency of the carrier is varied in accordance with the frequency and amplitude of the modulating signal.**

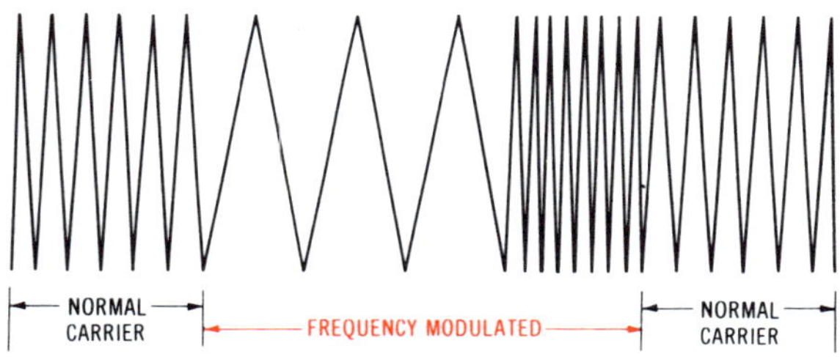

Fig. 7-6. Frequency modulation.

Notice in the diagram of Fig. 7-6 that the frequency of the fm carrier varies, but its amplitude remains the same. Now, study the diagrams in Fig. 7-7.

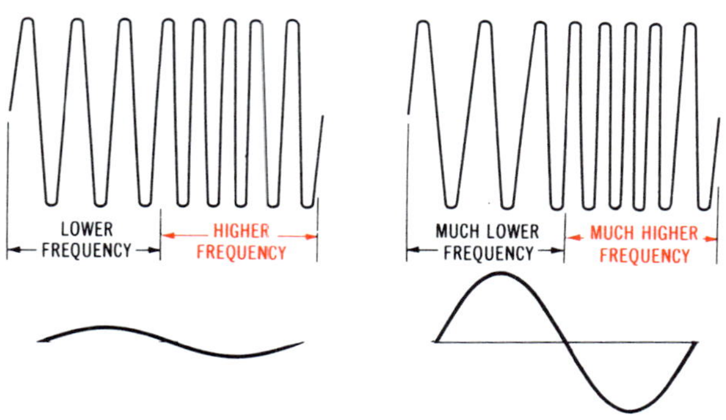

(A) Low-amplitude audio.

(B) High-amplitude audio.

Fig. 7-7. Carrier-frequency shift.

Q7-3. What is the difference between am and fm?

Your Answer Should Be:

A7-3. In am (amplitude modulation), the **amplitude** of the carrier is varied in accordance with the amplitude of the modulating signal. In fm (frequency modulation), the **frequency** of the carrier is varied in accordance with the frequency and the amplitude of the modulating signal.

In the examples illustrated by the two diagrams in Fig. 7-7, the frequency of the carrier is decreased by the positive portion of the audio signal and increased by the negative portion. The higher the amplitude of the modulating signal, the greater the shift in carrier frequency will be. However, the **average** carrier frequency during a complete cycle of modulation is equal to the normal frequency of the carrier because the shifts of carrier frequency balance each other.

Some rf signal generators generate a frequency-modulated signal. Other generators are designed with provisions for both amplitude and frequency modulation.

Output Circuit

Since the amplitude of the signal developed by most oscillators is relatively low, many signal generators include at least one stage of amplification. The mixing of generator and modulation oscillations usually takes place in this stage (Fig. 7-8).

A gain control is normally provided to vary the magnitude of the output signal. To permit impedance matching of the signal generator with the circuit to which the test signal is fed,

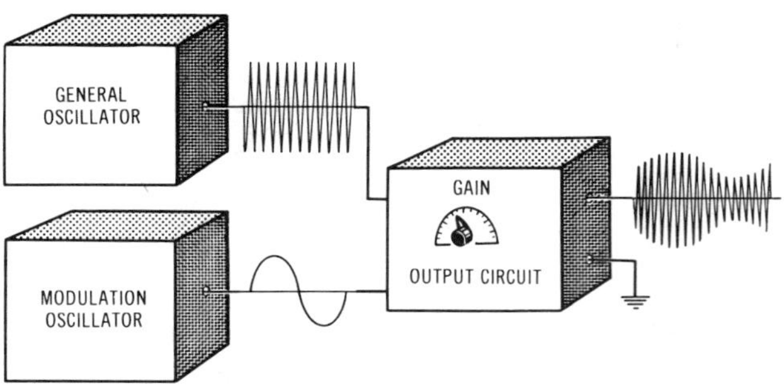

Fig. 7-8. Amplifying stage of a signal generator.

a cathode follower or similar impedance-matching stage is often employed. An attenuating network may be included to provide signals at a precise level of amplitude.

AUDIO-FREQUENCY SIGNAL GENERATOR

As previously indicated, an af signal generator covers the audible frequency range. Most af generators include a generator oscillator and an amplifier stage.

Requirements for Oscillation

A circuit must fulfill three requirements before it will oscillate (Fig. 7-9). It must have a means of **amplification,** a method of **feedback,** and some manner of **frequency control.**

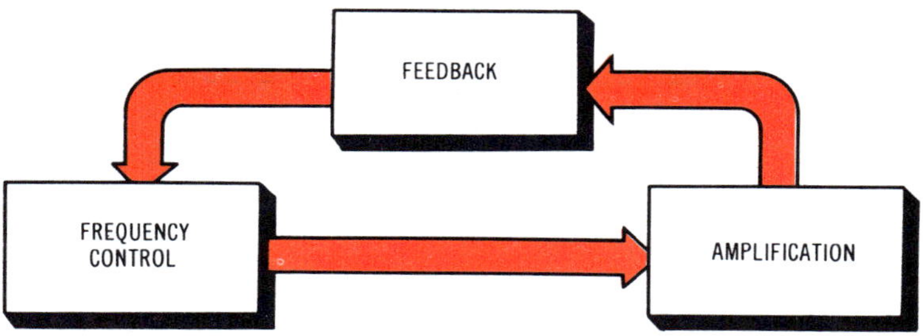

Fig. 7-9. Requirements for oscillation.

Amplification—An amplifier with collector current flowing will normally amplify any spurious signals appearing at the base. A spurious signal is any unwanted signal that is generated either in the equipment itself or externally.

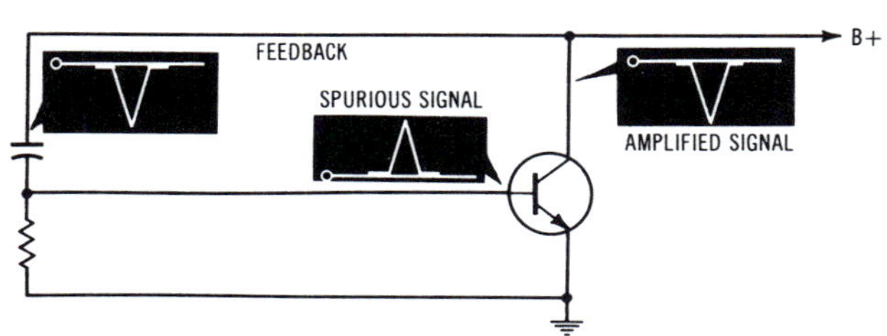

Fig. 7-10. Amplification and negative feedback.

Q7-4. What is the purpose of a gain control?

Q7-5. What are the circuit requirements for oscillation?

Your Answers Should Be:

A7-4. The gain control **varies the amplitude of the output signal.**

A7-5. An oscillating circuit must possess a means of **amplification, feedback,** and **frequency control.**

The signal appearing at the collector will be of the same shape as the base signal, but it will be larger in amplitude and reversed in phase by 180° (Fig. 7-10). If a negative-going signal from the collector were fed back to the base in addition to a positive-going input signal, the two signals would cancel. This process is called **negative feedback.** The signal returned from the collector must go through another 180° of phase reversal before it will be in phase with the base signal and in a position to aid amplification. This aiding feedback is called **positive feedback** (Fig. 7-11).

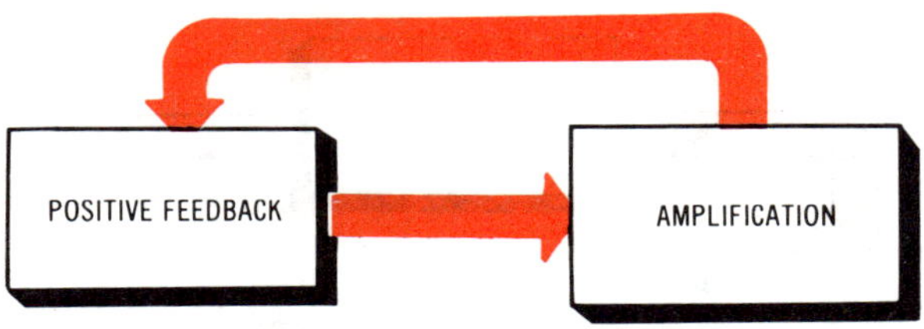

Fig. 7-11. Positive feedback.

Positive Feedback—Although there are several ways of accomplishing a second phase reversal of 180°, the basic principle is shown in Fig. 7-12. If the coils of the transformer between the collector and the base are wound so that they cause

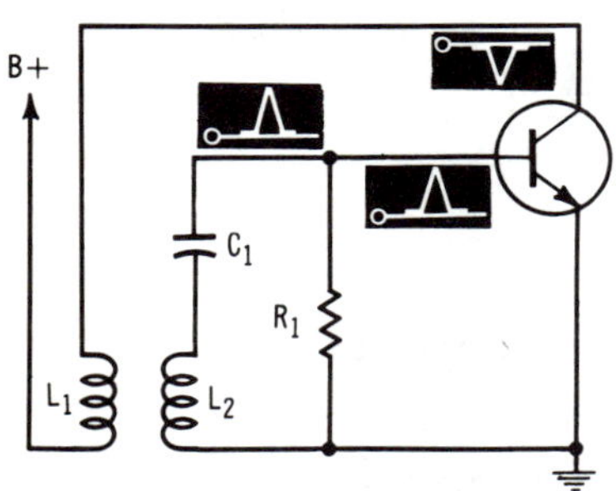

Fig. 7-12. Amplification and positive feedback.

a 180° phase inversion between the primary and secondary, the resultant signal will be in phase with the signal on the base. The amplitude of the signal will depend on the amount of voltage across L_1 and the resistive losses in the circuit. The remaining voltage is added to the base-to-emitter positive potential, causing an increase in collector current. The signal returning to the base on the next cycle will be increased proportionately. Waveforms in each succeeding cycle will be increased in amplitude in a similar manner. The signal amplitude will increase to a level established by the saturation point of the transistor.

You have seen how positive feedback aids in the build-up of a signal. However, something more must be added to the circuit before it will generate a steady signal of known frequency.

Frequency Control—Positive feedback will provide for a steady state of amplification and produce a constant-amplitude signal at the collector, but it cannot achieve a constant frequency. The circuit shown in Fig. 7-12 will amplify any spurious signal that appears at the base. The frequency of oscillation will be random. The third requirement of an oscillator—frequency control—now becomes apparent. First, how can frequency be regulated? How can one frequency of a group be selected for amplification in preference to the rest? The answer is the resonant circuit.

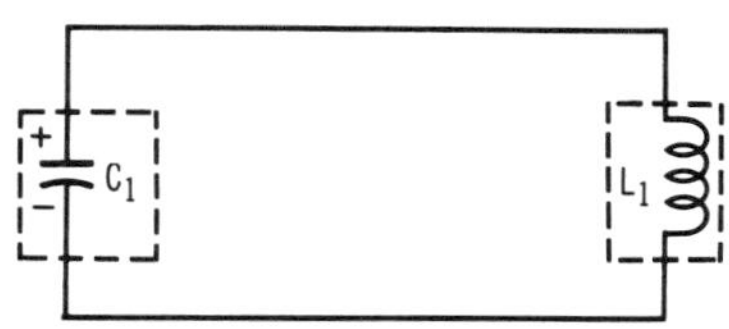

Fig. 7-13. Resonant circuit.

When the source is removed from capacitor C_1, the capacitor will begin to discharge. Current will flow through coil L_1, in

Q7-6. If a charge were placed on C_1 (shown in Fig. 7-13), in which direction would the capacitor discharge through the circuit—clockwise or counterclockwise?

Q7-7. Would current flow be maximum or minimum at the instant discharge begins?

Q7-8. What effect would L_1 have on the discharge current?

Your Answers Should Be:

A7-6. Discharge current would flow **counterclockwise** through the circuit—from the negative plate, through the coil, and back to the positive plate.

A7-7. At the instant discharge begins, current flow would be **maximum.**

A7-8. As the initial current surge flowed through L_1, it would produce an expanding magnetic field that would cause current in opposition to the surge. As the discharge current decayed, the field would collapse and cause a current in opposition to the decay of current. In effect, L_1 becomes a current generator to the degree that it **sustains the flow of current in one direction.**

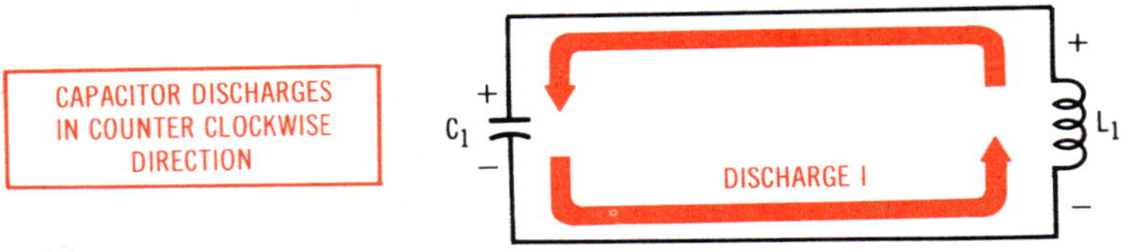

Fig. 7-14. Capacitor discharge.

Fig. 7-14, in the direction indicated and will cause a magnetic field to develop around the turns of coil L_1. When the charges on the plates of capacitor C_1 have equalized, current would normally stop flowing. However, the magnetic field around coil L_1 continues to collapse (initial collapse occurred as discharge current decayed), sustaining the current flow. Coil L_1

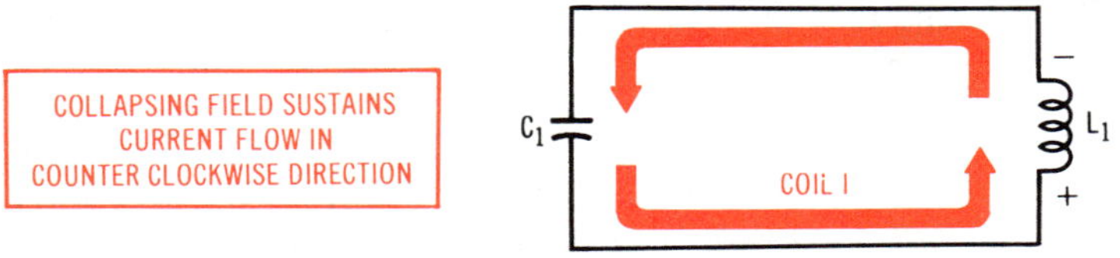

Fig. 7-15. Collapsing field of coil.

now becomes a voltage source, taking on the polarity shown in Fig. 7-15. Collapsing magnetic lines cut the coil turns, producing a current that builds an excess of electrons on the upper

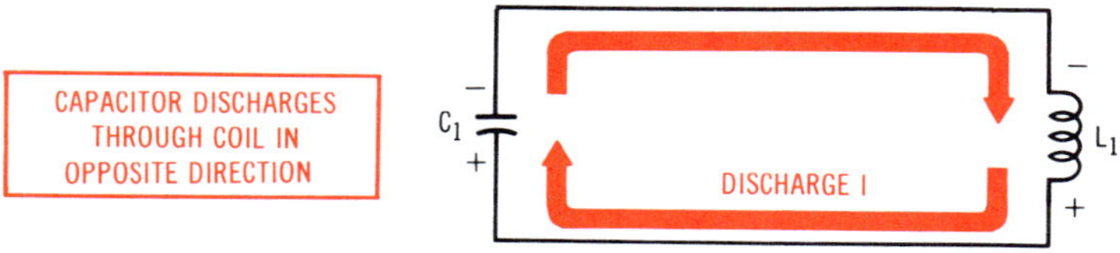

Fig. 7-16. Capacitor discharges through coil.

plate of C_1. Capacitor C_1 is charged in the opposite direction.

After the initial surge of current as capacitor C_1 discharges in the opposite direction, a magnetic field is developed to its maximum level across coil L_1. As discharge current begins to decrease, the field starts to collapse, developing a coil current in the same direction as the capacitor current. When the capacitor has discharged, the continuing collapse of the magnetic field will maintain current flow and will charge the opposite plate of the capacitor. The charge and discharge of capacitor C_1 will continue for several cycles.

The decreasing amplitude of current is often compared to two mechanical analogies. If you plotted the voltage across the capacitor in the previous example, you would obtain a graph resembling the one shown in Fig. 7-17.

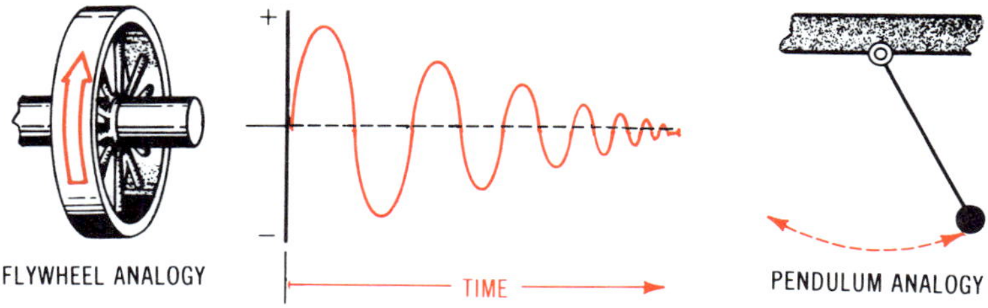

Fig. 7-17. Damping effect.

The LC circuit, the flywheel, and the pendulum have one quality in common—**damping effect.** Damping is the reduction of energy in a mechanical or electrical system as a result, in these cases, of absorption. The flywheel and the pendulum lose energy through absorption which is caused by the friction of the bearings and the surrounding air. The LC circuit (containing an unseen resistance R) gives up part of its energy in

Q7-9. What characteristic of the circuit prevents the cycling of current from continuing forever?

> **Your Answer Should Be:**
>
> **A7-9.** The **resistance** of the coil wire and circuit conductors dissipates some of the energy in the form of heat each time the current cycles through the circuit. Because of these losses, current flow will eventually decrease to zero. If there were no resistance in the circuit, the cycling would continue forever.

the form of heat as the current passes through the resistive parts of the circuit.

There is another analogy that explains the cycling action of the LC circuit. The momentum of the flywheel causes it to coast through several cycles of revolution. The momentum or potential energy built up by the pendulum as it completes each alternation of swing enables it to return through the next alternation to the other end of its swing. The corresponding effect in the LC circuit, of course, is the potential energy of the magnetic field that is developed by capacitor discharge current in passing through the coil. The collapsing field causes current flow to build up a charge on the capacitor in the opposite direction.

Self-Excited Oscillator

If an LC circuit were added as a frequency-control device to the other requirements of oscillation (feedback and amplification), the circuit should oscillate (Fig. 7-18). By adding a capacitor (C_2) across coil L_2, an oscillating circuit is developed that will be resonant at some specific frequency. Coil L_2

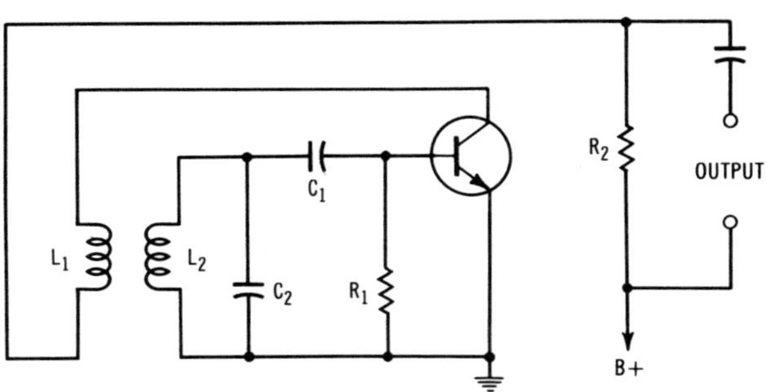

Fig. 7-18. Self-excited oscillator.

and capacitor C_2 form the resonant circuit. A signal in the collector circuit will be induced by L_1, called **a tickler coil,** into coil L_2. The voltage across coil L_2 will be in phase with the signal already on the base. The base voltage will be increased when it coincides with the instant of maximum voltage, plus-to-minus, top-to-bottom, across the resonant circuit. This will occur at a regular frequency determined by the values of coil L_2 and capacitor C_2; their size determines the charge and discharge time of the capacitor. As you recall, the resonant frequency of an LC circuit can be computed by:

$$\text{resonant frequency} = \frac{1}{2\pi\sqrt{LC}}$$

If coil L_2 were 1 millihenry in size and capacitor C_2 were 10 microfarads in size (both fairly large components), the resonant frequency of the LC circuit would be 1592 cycles per second (1592 hertz).

An LC circuit is often referred to as a **tank circuit.** The complete circuit, because of its tickler-coil feature, is called a **self-excited oscillator.**

The damping effect of the tank circuit is overcome by the induced voltage applied by the tickler coil in phase with the signal on the grid. The principle is much the same as applying a nudge to a pendulum that is sufficient to cause it to swing through the same distance of arc in each oscillation.

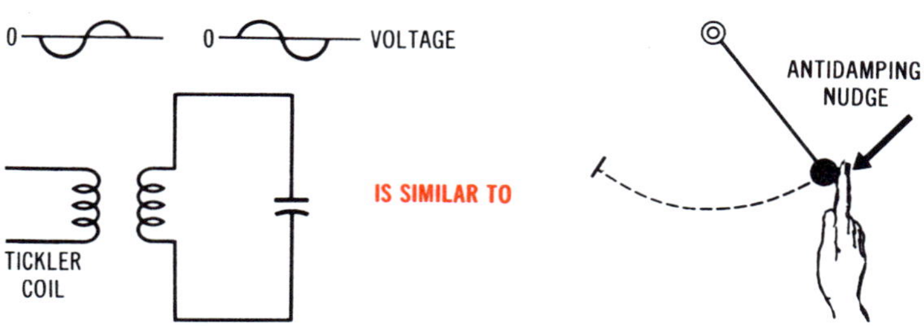

Fig. 7-19. Prevention of signal damping.

Q7-10. If either L and C, or both, are increased in value, what happens to the resonant frequency?

Q7-11. What is the resonant frequency of a tank circuit that has a capacitance of 25 pF and an inductance of 9 mH?

Your Answers Should Be:

A7-10. If L or C, or both, are increased in value, the resonant frequency of the oscillator will be **decreased.**

A7-11. The resonant frequency is **335.9 kHz.**

A constant frequency of tank-circuit oscillation is practical for only a few purposes. It can be used, for example, as a code practice oscillator for learning International Morse code or for testing circuits where a single frequency is adequate. To make the self-excited oscillator more versatile, either L or C must be capable of being changed. In most applications, C is a variable capacitor.

Phase-Shift Oscillator

Another circuit that is often used as an oscillating stage in an af signal generator is a **phase-shift oscillator.** The circuit shown in Fig. 7-20 uses a field-effect transistor.

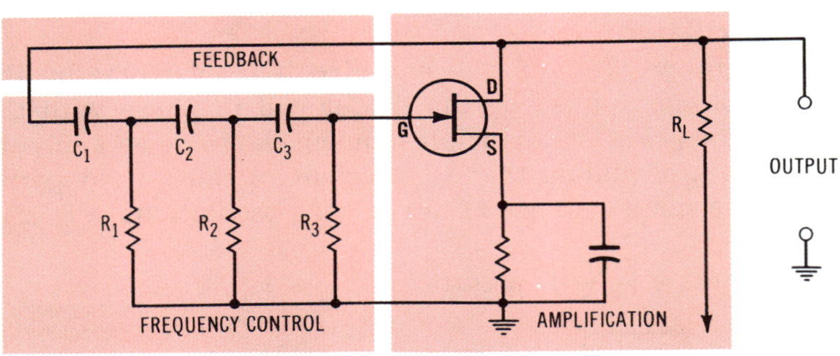

Fig. 7-20. FET phase-shift oscillator.

Notice that the frequency-control portion of the circuit does not contain a tank circuit. Instead, the frequency control uses three capacitors and three resistors. These RC combinations will not oscillate. They will, however, shift the phase of the feedback signal by an amount that is dependent on the ratio of R to C. If the values of each RC combination are selected to shift the signal 60°, the total shift of the returned signal at the gate of the FET will be 180°. This meets the basic requirement of frequency control, returning the signal in phase with the gate signal. The signal on the drain (D) has already been shifted 180°.

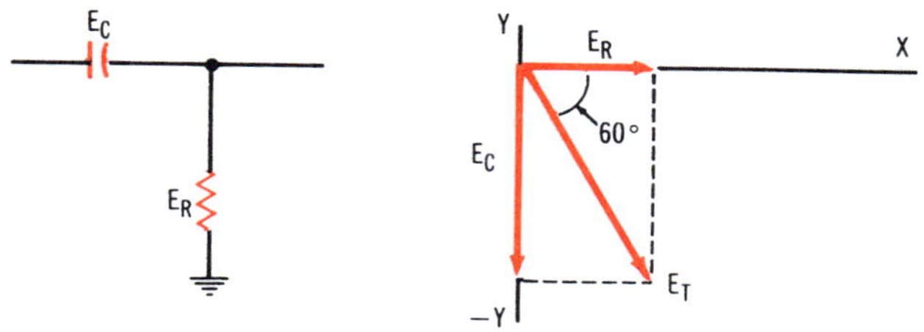

Fig. 7-21. A 60° phase shift.

Current through a capacitor will lead an ac voltage applied across the capacitor. If it is a pure capacitance with no resistance involved, the angle of lead will be 90°. Obviously, two capacitors (with no resistance), in series with the gate circuit, behave as one capacitor and will not achieve the required 180° phase shift. Resistance is required to develop the desired voltage between the gate and source. Although any number of RC combinations can be selected to accomplish the desired phase shift, a set of three is the best compromise. A 60° phase shift can be explained by the diagram of Fig. 7-21.

The values of C and R are selected so that the current in the circuit leads the applied voltage (E_T) by an angle of 60°. The voltage (E_R) across the resistor also leads the applied voltage by 60°; E_R is the output voltage of the RC combination. Three RC combinations produce a phase shift of 180°.

For a given value of C, there is only one frequency at which the phase shift of the RC combination is exactly 60°. The three capacitors in a typical phase-shift oscillator, as shown in Fig. 7-22, can be simultaneously varied to achieve a wide range of audio frequencies.

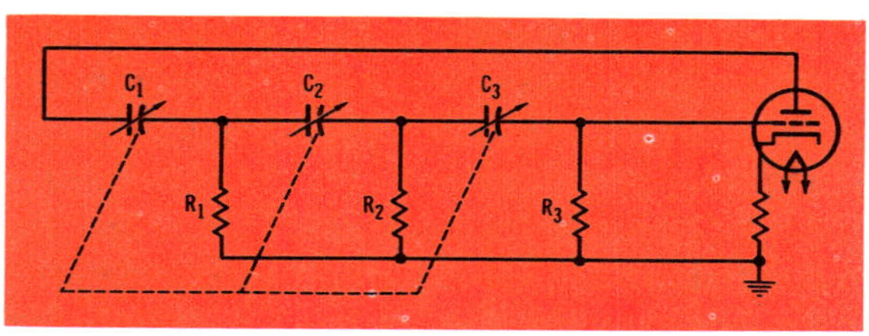

Fig. 7-22. Diagram for Problem Q7-12.

Q7-12. What characteristics of R and C, as shown in the schematic of Fig. 7-22, determine the frequency?

Your Answer Should Be:

A7-12. The **size of R or C** determines the frequency as well as the amount of time required for the capacitor to charge or discharge through the resistance. The longer the charge time, the lower the frequency will be.

Other Types of Oscillators

There are other types of oscillators used in an af frequency generator. Another popular type, operating on principles similar to the RC applications given in the phase-shift oscillator, is the **Wien-bridge oscillator.** A few signal generators, requiring only a single frequency output, use the accurate oscillating frequency of a **tuning fork.**

Typical Af Signal Generator

The block diagram of Fig. 7-23 represents a typical af signal generator. Usually the output circuitry contains an amplifier

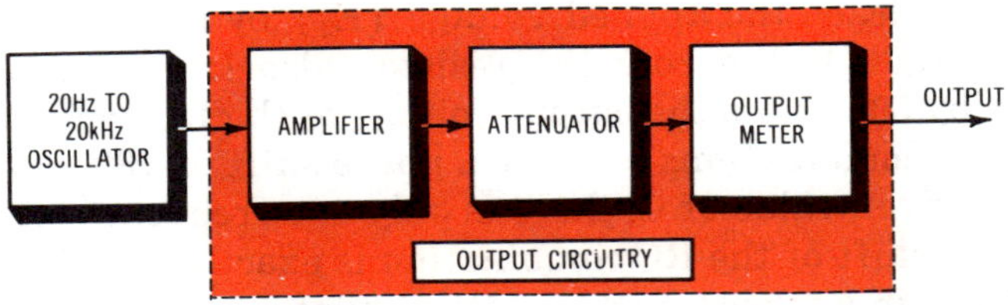

Fig. 7-23. Block diagram of an af signal generator.

with a network or attachment that permits matching the impedance of the external circuit to that of the generator. If the impedance mismatch is severe, the generator may be loaded down enough to change the oscillator frequency.

Some af signal generators have an output meter to indicate the amplitude of the generated signal. This feature is necessary when a signal of a specified amplitude is required for testing purposes—measuring the gain of a stage, for example. Attenuating networks composed of series and parallel resistances are used to supply signals at specified amplitudes. Some generators incorporate the features of both an output meter and attenuating networks.

USING AN AF SIGNAL GENERATOR

An audio-frequency signal generator (sometimes called an audio oscillator) has several useful test purposes. Within the range of its frequency coverage (generally 20 Hz to 20 kHz), it can be used as a signal source to check equipment or circuits designed to pass an audio signal (Fig. 7-24).

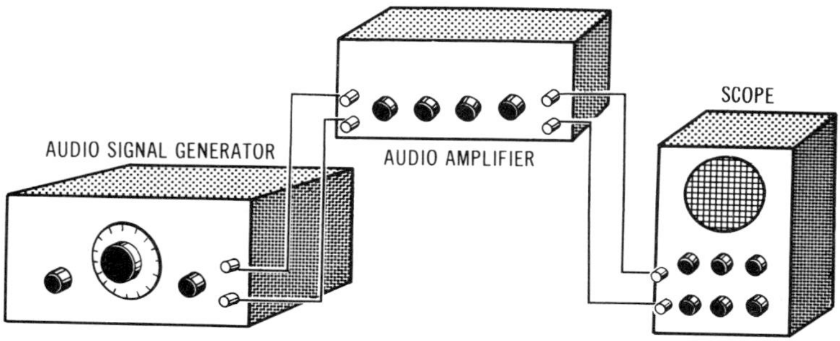

Fig. 7-24. An af signal generator used as a signal source.

Frequency Response and Fidelity

An audio amplifier used with a record player, hi-fi, or stereo system must have a good **frequency response** through the audible frequency range. Most people have an audible range of 15 and 15,000 hertz. Audio equipment with a flat frequency response between these limits is considered to be a faithful reproducer of sound. The word **fidelity** is used in reference to sound equipment. Good fidelity and flat frequency response both refer to the accuracy with which an input signal is reproduced at the output of a circuit or piece of equipment.

Q7-13. What are the main circuits in an audio-frequency signal generator?

Q7-14. Why must you be careful when matching the output impedance of a signal generator with that of the circuit being tested?

Q7-15. Why would an output meter be required in an af signal generator?

Q7-16. What is the normal frequency range of an af signal generator?

Q7-17. What are fidelity and flat frequency response?

Your Answers Should Be:

A7-13. **Oscillator, amplifier, attenuator,** and **output meter.**

A7-14. If they were badly mismatched in impedance, the signal generator would load down and could cause the oscillator to **change frequency.**

A7-15. The output meter is required to determine when a signal is of the **required amplitude** for a specific purpose.

A7-16. **20 to 20,000 Hz.**

A7-17. They are both a **measure of the accuracy** with which the input characteristics of a signal are reproduced at the output of a circuit.

Plotting Frequency Response in the Audio Range

Evaluation of fidelity or plotting a frequency-response curve can be accomplished by using the equipment setup shown in Fig. 7-24. If an oscilloscope is being used as the output indicator, it should have a graticule and be calibrated to measure ac voltage accurately.

Allow 15 minutes for the signal generator and amplifier to warm up. After each new frequency setting, plot the voltage reading on a graph as shown in Fig. 7-25. Generator settings for readings in the rising and falling portions of the curve should be made in multiples of 500 Hz. Intervals of 2000 Hz are satisfactory along the top part of the curve.

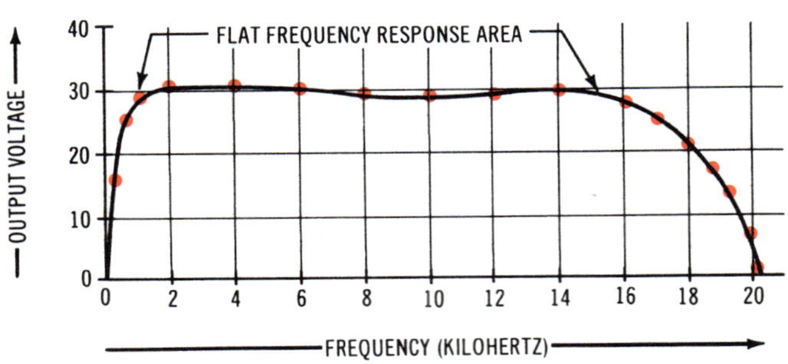

Fig. 7-25. Plotting frequency response.

A frequency-response curve is a line drawn through the dots plotted on the graph. It will reveal the fidelity of the amplifier for which it has been made. Do not adjust the gain of any equipment (generator, amplifier, or scope) during the test. If you do, output readings will be in error.

Determining Response at High Frequencies

A square wave has numerous odd harmonics. Depending on the steepness of the slope and the flatness of the peak, a square wave can contain harmonics with frequencies 10 to 100 times the frequency of the fundamental sine wave. Because of this quality, the square wave can be used to reveal frequency response when it is applied to the input of a circuit and its output is observed on a scope. Typical response indications are shown in Fig. 7-26.

Although commercial square-wave generators are available, there is a more economical way to develop a square wave for testing purposes. The sine wave from an af signal generator is fed into an amplifier operated in the class-A portion of the characteristic curve. Amplifier gain is increased until the tube or transistor is driven into saturation on the positive cycles and cut-off on the negative. In effect, this operation flattens out the positive and negative peaks of the sine wave. The output is a reasonable and suitable copy of a square wave.

Q7-18. Why should you not adjust the settings (other than frequency) of equipment during a frequency-response check?

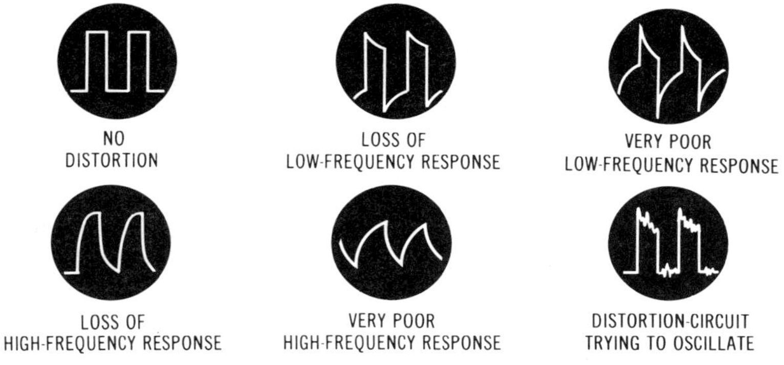

Fig. 7-26. Frequency response indications.

Your Answer Should Be:

A7-18. If gain or similar adjustments are made to equipment during a frequency-response check, **false amplitude readings** will be obtained.

FUNCTION GENERATORS

Function generators are a type of signal generator that can provide any of several output waveforms, with provisions for varying the frequency and amplitudes of each. Function generators, also called waveform generators in some versions, can provide, for example, outputs that are sine waves, triangle waves, sawtooth waves, positive square waves, and negative square waves. Special-purpose function generators can also develop other waveshapes. The output waveforms are shown in Fig. 7-27.

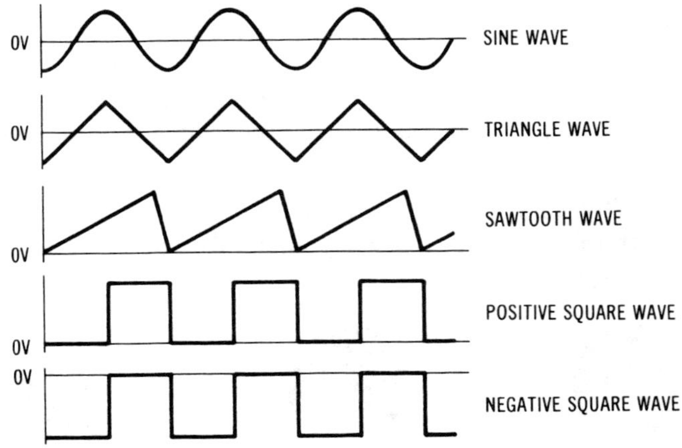

Fig. 7-27. Output waveforms from a function generator.

Some function generators include **sweep control** capability. That means that the output waveform can be automatically varied in frequency over a selected range.

One example of using a function generator is shown in Fig. 7-28 where the square-wave output of a generator is connected to the input of an audio amplifier. A square wave is made up

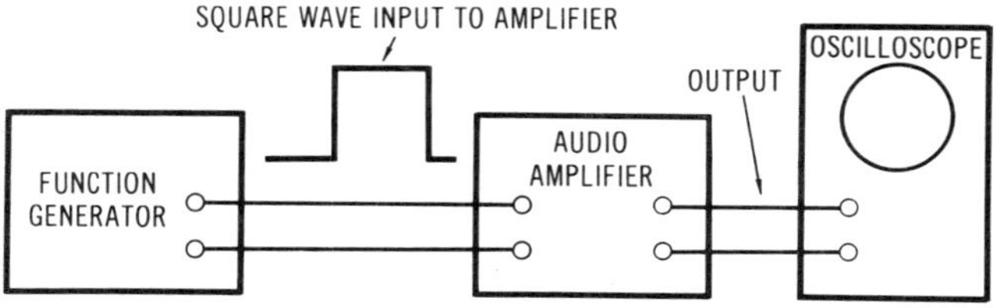

Fig. 7-28. Hook-up for using a function generator.

of a number of odd harmonics of the fundamental frequency—usually up to 20 times the fundamental. An oscilloscope is used to observe the resulting square wave at the output of the amplifier. The effects of different types of amplifier distortion on the output waveform are shown in Fig. 7-29. The oscilloscope is preferably a dual-trace or dual-beam type so that the input wave and output wave can be conveniently compared.

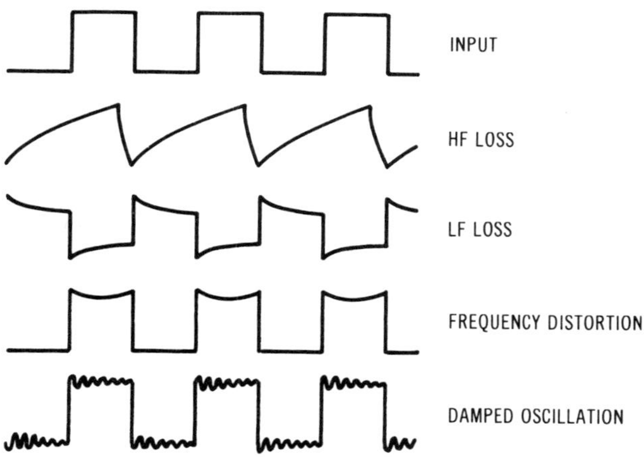

Fig. 7-29. Distortion of the output waveforms.

Q7-19. **Sine waves, triangle waves, sawtooth waves, and square waves are examples of signals that can be obtained from a _ _ _ _ _ _ _ _ _ generator.**

Q7-20. **The term "sweep control" as applied to a function generator means that the output signal can be varied in _ _ _ _ _ _ _ _ _ _ over a selected range.**

Your Answers Should Be:

A7-19. Sine waves, triangle waves, sawtooth waves, and square waves are examples of signals that can be obtained from a **function** generator.

A7-20. The term "sweep control" as applied to a function generator means that the output signal can be varied in **frequency** over a selected range.

RADIO-FREQUENCY SIGNAL GENERATOR

The rf signal generator is identical in principle to the af generator except that the rf signal generator can develop frequencies up to several hundred megahertz.

Typical Rf Signal Generator

The diagram of Fig. 7-30 shows a simplified block diagram of a typical rf signal generator. Notice that the rf and audio

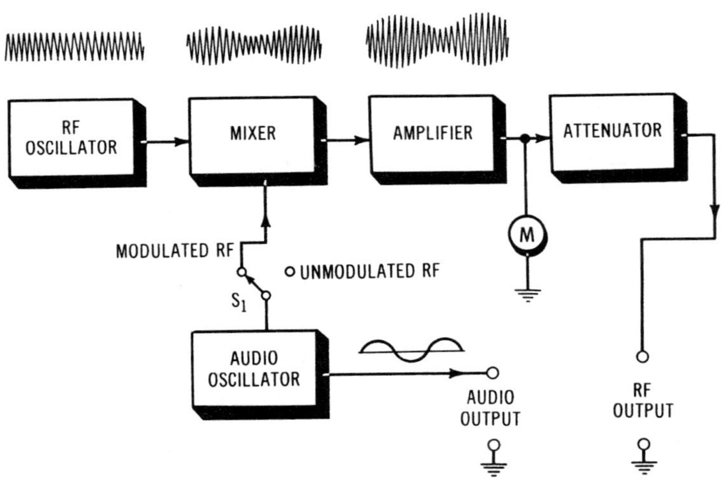

Fig. 7-30. Block diagram of an rf signal generator.

oscillators are connected to a mixer stage when switch S_1 is in the MODULATED RF position. The mixer circuit superimposes the audio on the rf carrier and feeds the modulated wave to the output through the amplifier, attenuator, and meter. When switch S_1 is in the UNMODULATED RF position, a pure carrier frequency appears at the RF OUTPUT and an audio signal is fed to the AUDIO OUTPUT.

Rf Oscillator—There are several types of oscillator circuits that can be used as rf generators. A widely used circuit for this purpose is the **Hartley oscillator.**

Coils L_1, L_2, and capacitor C_2 form the tank circuit. The collector signal is fed through the dc blocking capacitor (C_1) to the bottom of coil L_2. Coil L_2 induces a voltage in L_1 that is 180° out of phase with the collector signal (in phase with the

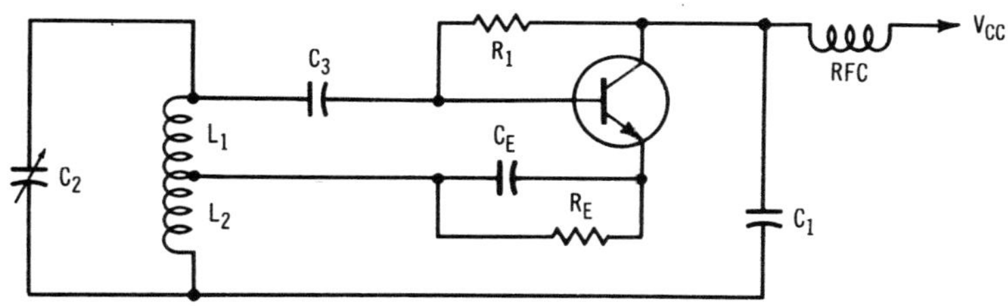

Fig. 7-31. Hartley oscillator.

base signal). The frequency of the tank is adjusted by C_2. The RFC (radio-frequency choke) in the V_{cc} line prevents ac on the collector from entering the V_{cc} source. Resistor R_1 establishes base bias. Solid-state Hartley oscillators operate similar to the vacuum-tube type just described.

Audio Oscillator—The major purpose of the audio oscillator is to modulate the carrier at an audible rate.

Buffer and Mixer Stage—In the circuit shown in Fig. 7-32, the two signals are applied across a resistance network which feeds the resultant voltage to the grid of the next stage.

The buffer places a stage of amplification between the rf oscillator and the output, thereby isolating the oscillator from the loading effects of an external circuit. The buffer also sepa-

Q7-21. What is the major difference between an af and an rf signal generator?

Q7-22. What are the main circuits in the rf signal generator?

Q7-23. What is the purpose of the mixing stage?

Your Answers Should Be:

A7-21. An af signal generator can usually produce frequencies up to **20,000 hertz,** while some rf signal generators can develop frequencies up to **several hundred megahertz.**

A7-22. The **rf oscillator, mixer, af oscillator, amplifier,** and **attenuator.** An output meter is often included, but it is not a necessity.

A7-23. The mixer stage joins the carrier and audio frequencies **to produce a modulated carrier.**

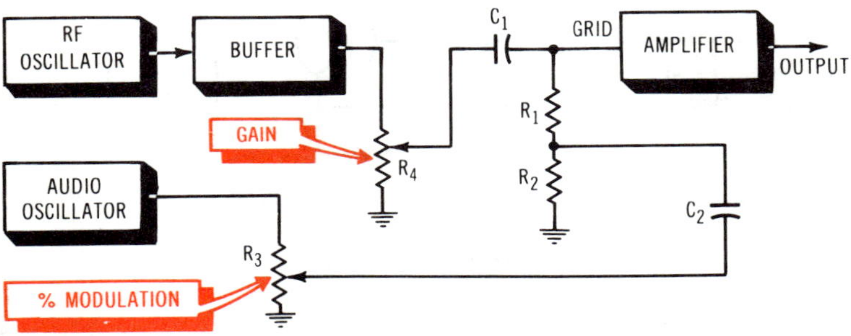

Fig. 7-32. A method of mixing signals.

rates the rf oscillator from the modulating voltages of the audio oscillator. The carrier output from the buffer is across R_4 (marked GAIN on the front panel), a variable resistance. The control may have other names, but its purpose is to regulate the amplitude of the carrier. Voltage is taken off resis-

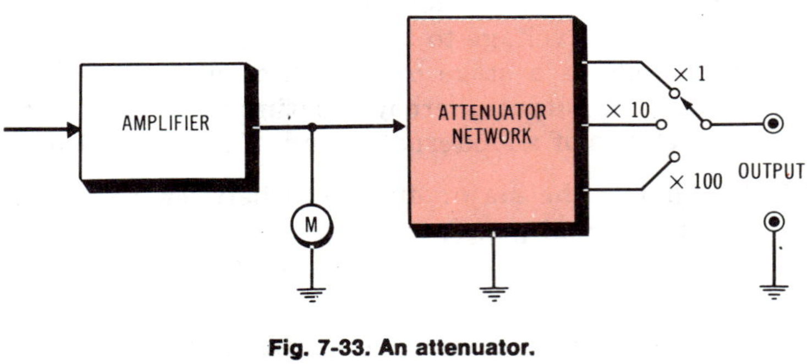

Fig. 7-33. An attenuator.

tance R_4 and distributed across resistors R_1 and R_2 and, then, fed to the grid of the amplifier.

The output of the audio oscillator appears across R_3, which is adjusted to select the desired percentage of modulation. This voltage is applied to the junction of resistors R_1 and R_2. Voltage, changing at an audio rate across resistor R_2, is subtracted from or added to the carrier amplitude at the grid of the amplifier.

Attenuator—Attenuators are resistive networks that permit selection of reasonably precise voltages as outputs. The meter reading is usually multiplied by the indicated number to obtain the output amplitude of the waveform.

USING AN RF SIGNAL GENERATOR

The most significant application of an rf signal generator is its use in tuning, aligning, and troubleshooting radio receivers and other equipment.

Superheterodyne Receiver

A **superhet** (common name for superheterodyne) receiver is manufactured in two general varieties. One has a transformer power supply and the other does not. The latter is often referred to as an ac-dc set. A block diagram for a typical superhet is shown in Fig. 7-34.

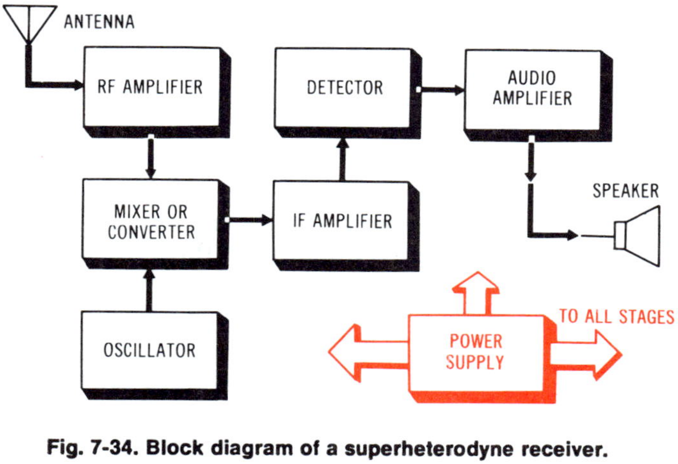

Fig. 7-34. Block diagram of a superheterodyne receiver.

The purpose of the radio receiver is to amplify the weak signal of a selected rf carrier and its audio modulation, remove the audio component from the carrier, and then amplify the audio so that it can be heard clearly on a speaker.

Antenna—The antenna is usually a loop or flat coil of wire attached to the back of the cabinet; it can be several turns of wire wrapped around an insulated rod. The antenna picks up carrier signals from transmitters within range of the receiver and feeds a selected signal to the first stage.

Rf Amplifier—In some sets, the antenna, in addition to picking up the signal, serves as a coil to form a resonant circuit with a capacitor. In other sets, a separate coil is used for this

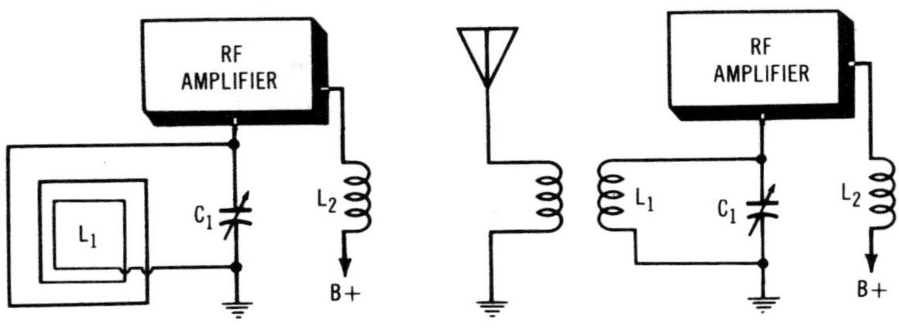

(A) Using the antenna as the resonant circuit coil.

(B) Using a separate coil as the resonant circuit coil.

Fig. 7-35. The antenna circuit.

purpose (Fig. 7-35). By varying the capacitor, L_1 and C_1 can be made resonant to a specific carrier frequency among the many appearing at the antenna. The selected frequency develops its voltage across the tank. The voltage is fed to the input of the rf amplifier. The amplified signal (carrier plus audio modulation) is applied across coil L_2, the primary of a transformer connecting the amplifier to the next stage. In some receivers that do not have an rf amplifier, the antenna coil is fed to the mixer stage.

Oscillator—This stage is designed in accordance with the basic principles of any oscillator—a stage of amplification, a means of signal feedback, and a method of frequency control. In this case, frequency is controlled by a tuned resonant circuit similar to the type used in the rf amplifier.

Mixer—Carrier and local-oscillator frequencies are heterodyned together in the mixer. **Heterodyning** is the mixing together of two signals to produce two additional frequencies which are the sum and difference of the originals. The purpose is to develop an **intermediate frequency**—the difference between the rf-carrier and the local-oscillator frequencies. If the intermediate frequency (if) is the same for all broadcast frequencies, the if amplifier and its frequency-selecting networks need only be capable of passing and amplifying just the one frequency.

This is the task accomplished by the mixer. It **beats** (another word for heterodyning) the oscillator and rf frequencies together and passes a total of four frequencies to the plate—carrier frequency, local-oscillator frequency, a sum frequency (carrier plus oscillator), and a difference frequency (oscillator minus carrier). In most sets, the value of the difference frequency is 455 or 456 kHz. It is obtained by tuning the oscilla-

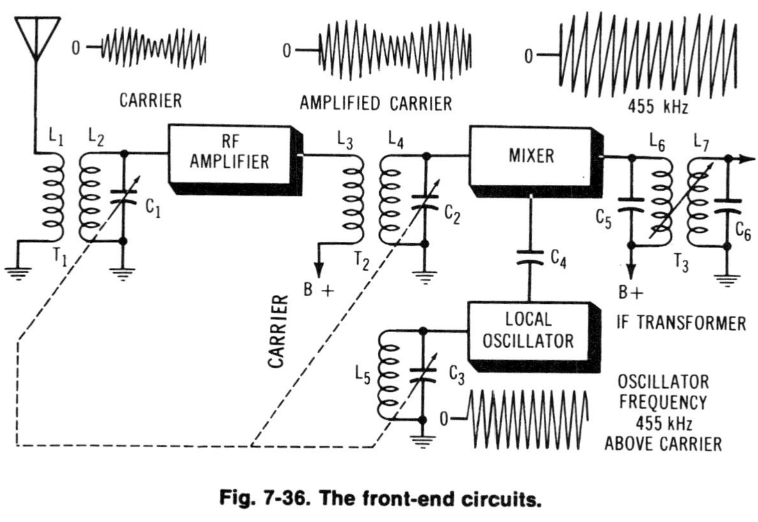

Fig. 7-36. The front-end circuits.

Q7-24. What is the purpose of the gain control on the rf signal generator?

Q7-25. What is the rf signal generator used for most frequently?

Q7-26. What is heterodyning?

Your Answers Should Be:

A7-24. The gain control regulates the **amplitude** of the carrier signal.

A7-25. **Tuning, aligning,** and **troubleshooting** equipment are three of the main uses for the rf signal generator.

A7-26. Heterodyning is the process of **mixing** two frequencies together to produce two other frequencies equal to the sum and difference of the first two frequencies.

tor tank circuit to a value always 455 or 456 kHz above any setting of the carrier-selecting circuits.

As can be seen from the diagram in Fig. 7-36, capacitors C_1, C_2, and C_3 are ganged together to a single tuning control. Capacitors C_1 (rf amplifier) and C_2 (mixer) are tuned to the desired carrier frequency, rejecting all other carriers. Capacitor C_3 (local oscillator) is tuned to exactly 455 kHz (the if frequency that will be used in the rest of this chapter) above the carrier; this is true no matter which station is selected. Of the four frequencies that appear at the plate of the mixer, only the difference frequency (455 kHz) will be selected by the if transformer.

Converter—The distinction between a mixer and a converter depends on whether or not the local oscillator employs a separate tube. If it does not, as is the case in most ac-dc superhets, the mixing tube acts as a mixer for the radio and as a means of amplification for the oscillator tank circuit (Fig. 7-37).

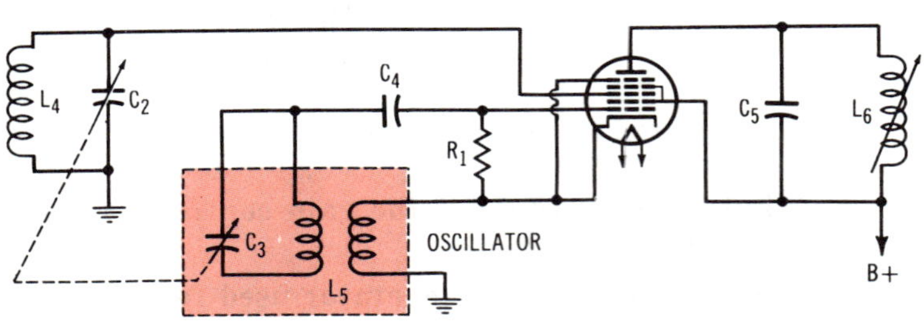

Fig. 7-37. A converter.

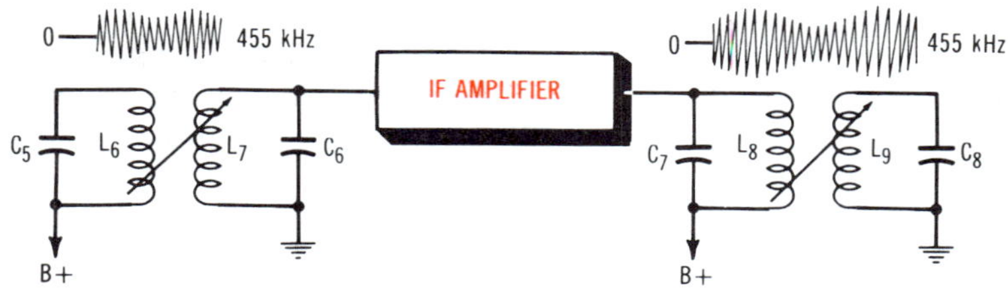

Fig. 7-38. The if strip.

***If Amplifier*—**The if stage consists of two tunable transformers separated by an amplifier (Fig. 7-38). More sensitive receivers may have two amplifiers and three transformers. Each of the coils in the transformers has a metallic core that can be adjusted to give the value of inductance necessary to make the respective tank circuits resonant at 455 kHz. By this method, only the difference frequency from the mixer or converter will be allowed to enter the if strip. The **if strip** is a common term that includes the if-amplifier tubes and transformers.

***Detector*—**The two small plates and the cathode of the tube, in Fig. 7-39, act as a diode that passes only the positive parts of the if signal. The time constant of resistor R_2 and capacitor C_9 is long enough that the voltage across resistor R_2 follows the envelope of the carrier rather than the individual half cycles. Thus, the audio signal is removed from the if carrier.

Fig. 7-39. Detector.

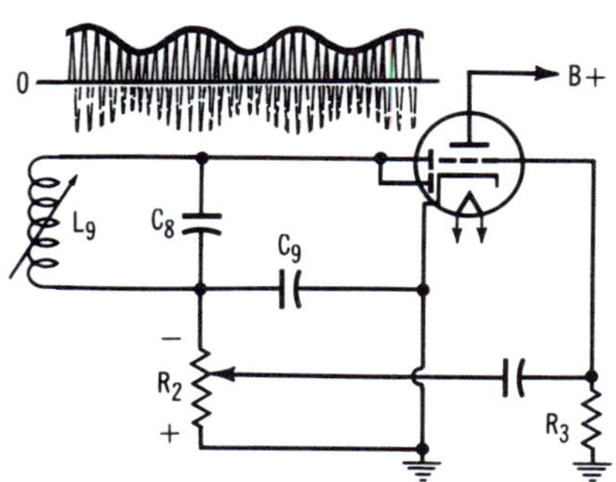

***Audio Amplifier*—**This stage further amplifies the audio signal and feeds it to the primary of the speaker transformer. The speaker is energized, and the audio originating at the broadcast studio is reproduced.

Automatic Volume Control (avc)—Avc is a feedback circuit that takes the average value of the rectified voltage (audio component) from the detector and feeds this average voltage back to the grids of the preceding tubes. This feedback voltage tends to change the bias on the tubes. The gain is therefore changed to compensate for changes in received signal voltage. (See Fig. 7-40.)

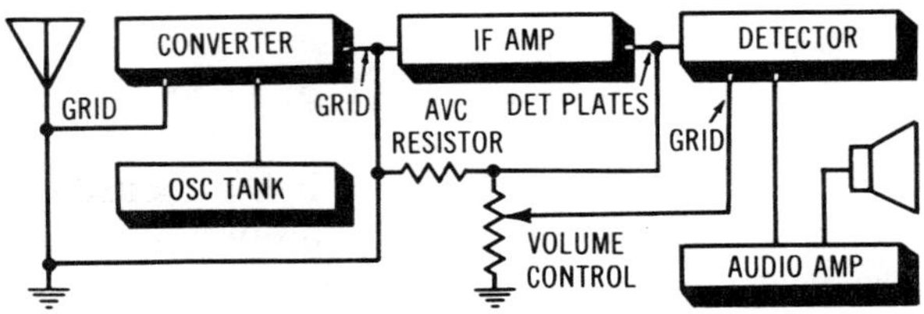

Fig. 7-40. An ac-dc superhet.

Solid-State Front-End Circuits

A solid-state circuit for the "front end" of an fm receiver is shown in Fig. 7-41. It includes the local oscillator and mixer. The output of the transistor oscillator stage is capacitively coupled through C_{13} to one of the gates of the dual-gate-protected MOS field-effect transistor mixer stage. A similar MOS device precedes the mixer and operates as an rf amplifier providing preamplification before mixing and, also, a greater selectivity than if the rf stage were not used. The other gate of the MOS transistor receives the output of the rf stage. The use of separate gates and the rf stage also helps to reduce oscillator radiation.

Fig. 7-42 shows an integrated circuit as it is used in an automobile receiver. In this design, tuning is accomplished with coils that are mechanically ganged together to vary the frequency of the rf amplifier, the antenna circuit, and the local oscillator. The if frequency is 262 kHz, a frequency that is commonly used for automobile radio receivers. The IC used in Fig. 7-42 is the RCA CA3123E AM Radio Receiver Subsystem. A connection diagram of such a device, along with its pin designations, is shown in Fig. 7-43. The IC contains all the electronic elements needed for rf amplification, mixing, oscillator generation, if amplification, agc detection, and voltage regu-

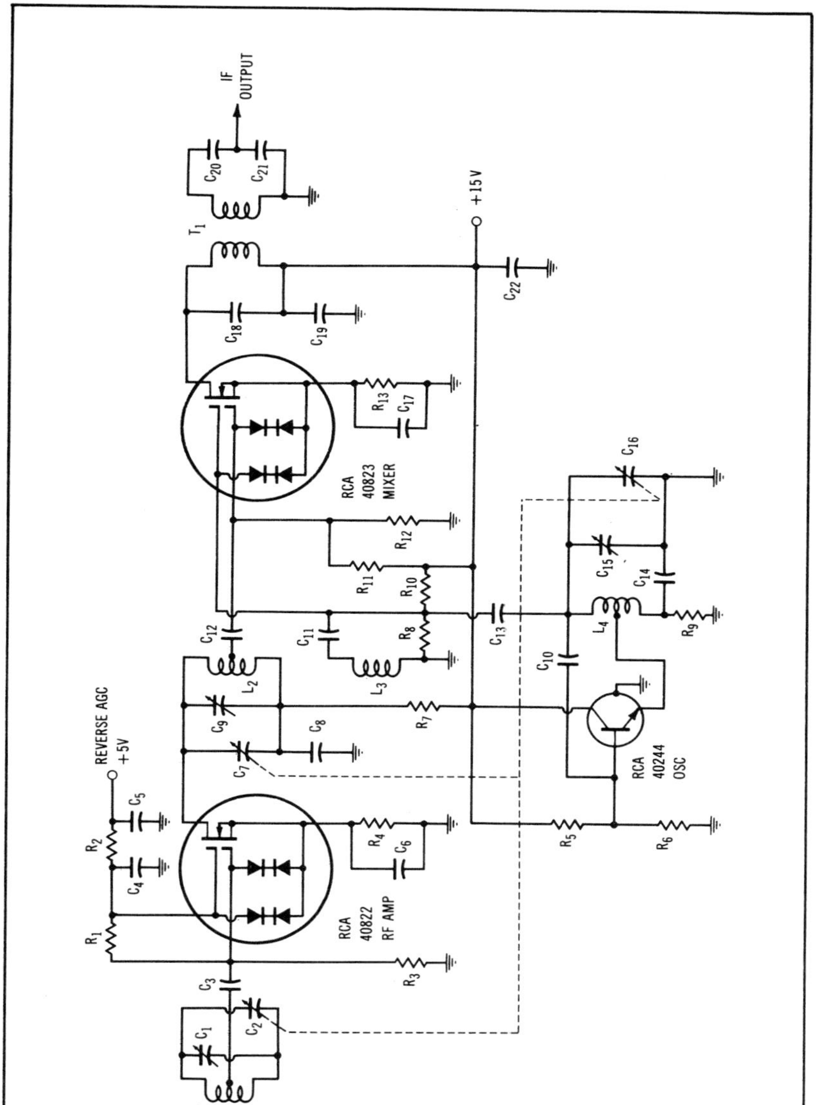

Fig. 7-41. The "front end" of an fm receiver. (*Courtesy RCA.*)

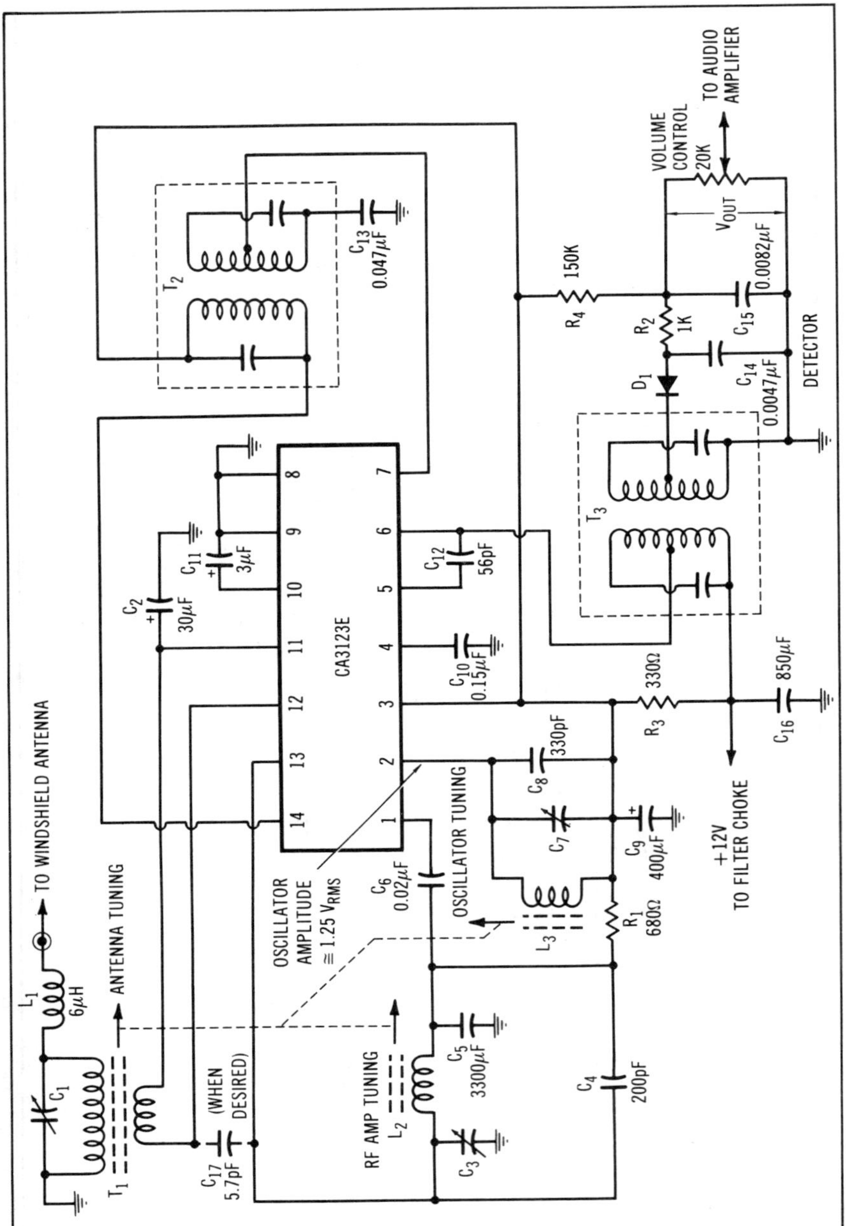

Fig. 7-42. Using an RCA CA3123E in an automobile receiver. (*Courtesy RCA.*)

lation. Only the tuning capacitor and its operating mechanism are missing.

Present-day radio receivers often employ integrated circuits for the converter and mixer stages. ICs are also available that contain virtually all the electronic functions for a complete radio receiver, with the exception of the speaker, the on-off-volume controls, and the mechanical parts of the tuner. An integrated circuit of this type is typically called a "radio on a chip."

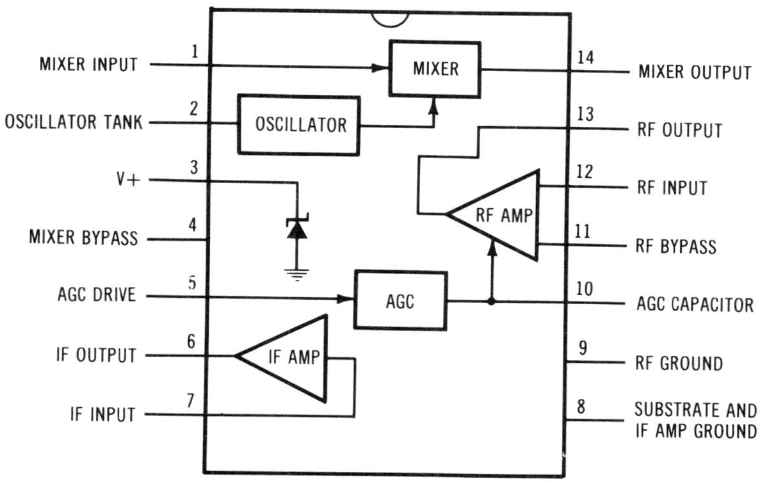

Fig. 7-43. An am radio receiver subsystem. (*Courtesy RCA.*)

Q7-27. Among the many carrier frequencies that appear on the antenna, how does a superhet select a single frequency?

Q7-28. What is the difference between a mixer and a converter stage in a superhet receiver?

Q7-29. The output of the transistor oscillator of the MOS mixer, shown in Fig. 7-41, is coupled through capacitor C_{13} to one of the _____ of the MOS mixer stage.

Q7-30. Use of separate input gates in the MOS mixer reduces the __________ from the local oscillator.

Your Answers Should Be:

A7-27. Capacitor C_1 and coil L_2 of the rf transformer (T_1) and capacitor C_2 and coil L_4 of the mixer transformer (T_2) are **resonant circuits** that can be tuned to a specific frequency, selecting the one desired carrier.

A7-28. A mixer **heterodynes,** or beats, the carrier and local-oscillator frequencies together to obtain two other frequencies, called the sum and difference frequencies. The converter stage performs the functions of the **mixer and oscillator** in a one-tube stage.

A7-29. The output of the transistor oscillator of the MOS mixer, shown in Fig. 7-41, is coupled through C_{13} to one of the **gates** of the MOS mixer stage.

A7-30. Use of separate input gates in the MOS mixer reduces **radiation** from the local oscillator.

Superhet Alignment

Although you may have enough background knowledge to undertake the task of receiver alignment, a word of caution is necessary. Make certain that you understand how your signal generator operates and how it is to be adjusted for use. Also, be sure you know how the receiver you are going to align works.

Why Should a Receiver Be Aligned?—The operating characteristics of tubes and other parts change with age. Periodic heating and cooling of parts is largely responsible for these changes. The change may not be great enough to make the set fail to operate, but it may be enough to throw the set out of alignment. In fact, most home receivers require at least a slight realignment. Improper alignment results in a lower receiver output, poor separation of stations, and a decrease in tone quality.

You should work from the schematic diagram of a receiver to obtain the information needed for alignment. Unfortunately, these diagrams are not always provided with commercial equipment. However, printed service information is available at most local electronic-parts supply houses. The service data usually contains a clearly drawn schematic for the equipment specified, showing the values of all parts and the voltages that should be read at the significant test points. A parts list, the location of alignment test points, and photographs showing the location of all parts are usually included.

Items Required for Realignment—The bare minimum of tools is shown in the illustration of Fig. 7-44. There are six

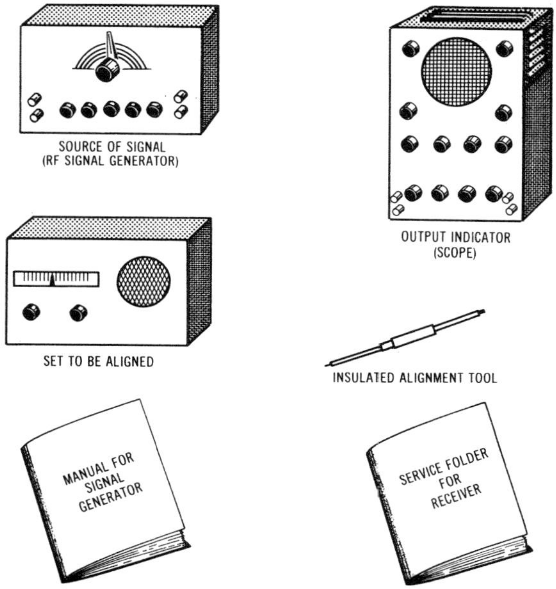

Fig. 7-44. Realignment devices.

items shown, and each is important. A few other common hand tools are also needed to remove the receiver from the cabinet. An insulated screwdriver is used to prevent the shorting or detuning of circuits while they are being adjusted.

Q7-31. What are the functions of the following tools when realigning a receiver?

(A) Rf signal generator.
(B) Output indicator.
(C) Operating manual for your signal generator.
(D) Service folder for the receiver.
(E) Common-sense application of technical knowledge.

Your Answers Should Be:

A7-31. (A) **Provides a carrier frequency** under controlled conditions.

(B) **Permits the observation of the results of adjustments** made to produce best performance.

(C) **Provides the proper operating procedures.**

(D) **Provides a schematic diagram and other information** necessary for alignment.

(E) **Without a common-sense application** of your technical knowledge, many **unnecessary mistakes will be made.**

Calibrating the Generator

Although most generators are designed to give good accuracy, it is always best to calibrate the output of the generator with a broadcast station. After the signal generator has been

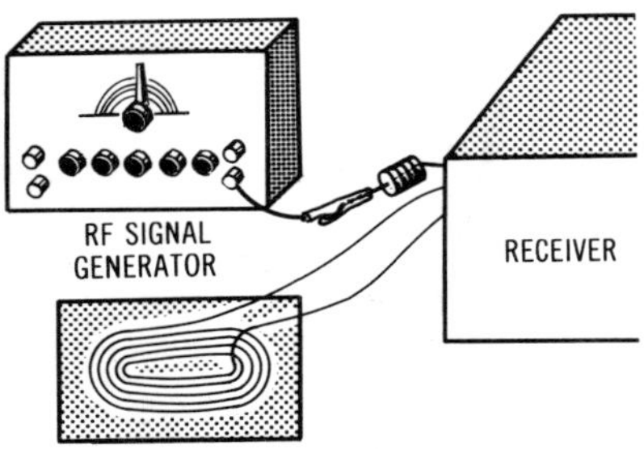

Fig. 7-45. Calibrating the signal generator.

on for about 15 minutes and the receiver has been tuned to a strong local station, hold the signal generator test clip or probe near the antenna terminal of the receiver. Adjust the signal generator dial approximately to the station frequency. If a tone or a squeal cannot be heard, connect the probe to the antenna terminal through a small-value capacitor. A noise should be heard on either side of **zero beat** (a point of silence). When the generator dial is at zero beat, the generator frequency is the same as the station frequency. Without retuning the generator, adjust the generator dial to read this frequency. (Consult the generator instruction manual.)

Alignment Steps

Receiver alignment consists of two steps. The if circuits are first tuned to 455 kHz, or whatever if frequency the manufacturer specifies. Then the rf, mixer, and oscillator circuits are tuned, in that order, for maximum signal. Connect the generator as shown in Fig. 7-46.

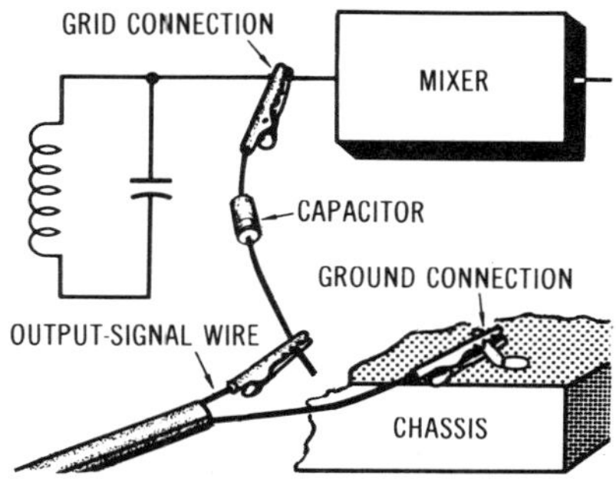

Fig. 7-46. Connections for tuning an if strip.

There is a wide choice of output indicators available. Whichever you choose, adjustments are made for maximum output indication and not for a precise numerical reading (Fig. 4-47).

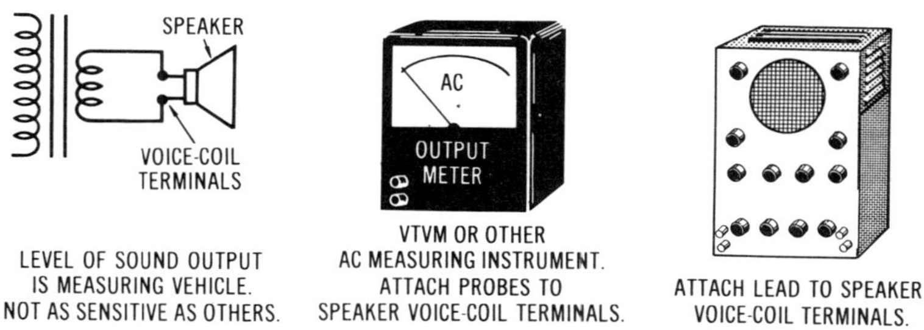

Fig. 7-47. Output indicators.

Set the signal-generator frequency to the if frequency of the receiver, and modulate the carrier with an audio signal. Set the receiver volume control to maximum gain. Ground the avc to the chassis for best output performance. Ground the grid of the local-oscillator tube, if the set has one (or follow the di-

Q7-32. What steps are required to align a receiver?

Your Answer Should Be:

A7-32. First, the **if circuits must be tuned to the specified if frequency** (normally, 455 kHz). Then, the **rf, mixer, and oscillator circuits are tuned for maximum signal.**

rections of the manufacturer, especially for solid-state receivers), to prevent any undesired signals from entering the if strip.

Tuning is accomplished by adjusting each of the two screws inside the top of the if can. Be sure to use an insulated screwdriver for this operation. (Solid-state receivers use transformers with powered-iron cores. They are tuned by adjusting the cores with an alignment tool.)

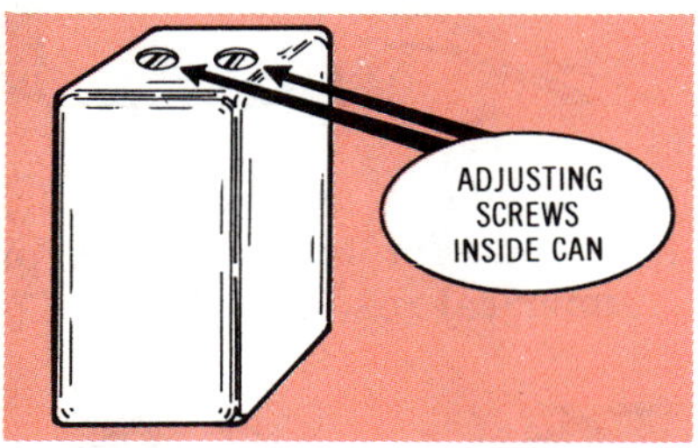

Fig. 7-48. An if can (transformer).

Carefully adjust the screws or cores for maximum reading on the output indicator. Start with the coil nearest the detector and adjust each, in turn, working toward the mixer.

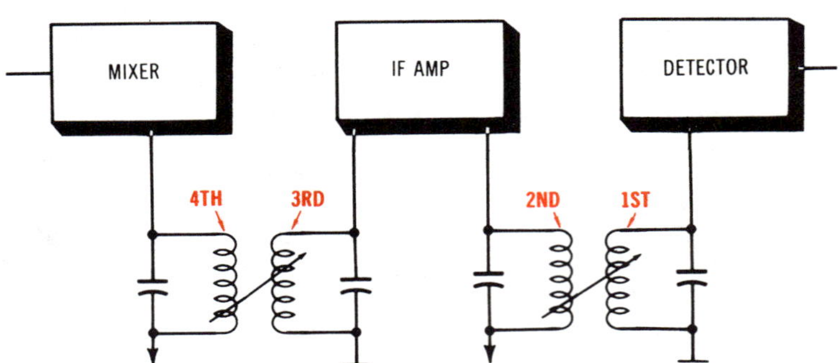

Fig. 7-49. The if tuning sequence.

During this and all following alignment steps, keep the gain of the signal generator no higher than necessary to give an indication of the output indicator used. If the gain is too high, receiver stages will overload or saturate, and output readings will be invalid. Readjust all coils until you are sure they have been peaked to maximum.

After the if coils have been peaked for a maximum output indication, the mixer and oscillator tuning circuits also must be adjusted to give a maximum-signal indication. **Front end** is a term applied to these tuning stages. There are some sets that have an rf amplifier; the tuning circuit of the rf amplifier must be adjusted in a similar manner.

To tune the front end, attach the generator high-side lead with its capacitor to the antenna terminal. Restore the oscillator to operational condition if you have previously disabled it. Tuning procedures involve adjusting both the small **trimmer** capacitors attached to the main tuning capacitor and the **padder** capacitor located in the oscillator tank circuit.

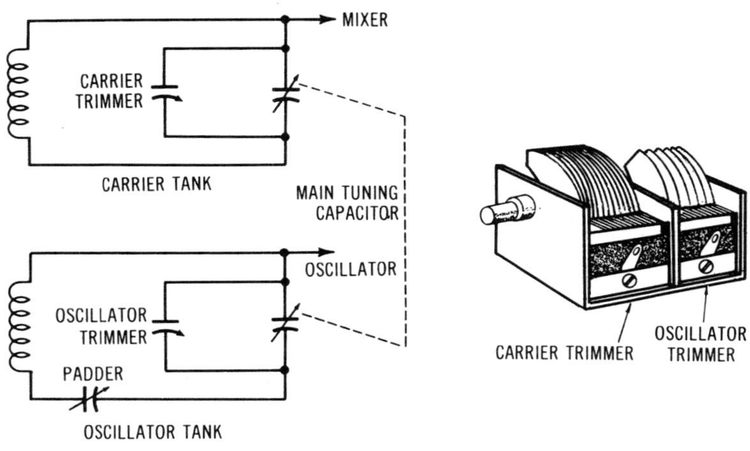

Fig. 7-50. Tuning capacitor.

Q7-33. At what end of the receiver should the first if adjustment be made?

Q7-34. During the tuning of the if coils, at what frequency and amplitude should the signal generator always be set?

Q7-35. At what point in the receiver should the generator leads be connected while adjusting the if coils?

Your Answers Should Be:

A7-33. The if coil to be adjusted first is the one **nearest the detector stage.**

A7-34. The signal generator should be set **at the if of the receiver,** modulated with an audio tone, and set at the **lowest amplitude that gives an output-indicator reading.**

A7-35. The high-side generator lead should be connected to the **mixer grid** or **the input terminal of the transistor** (or as otherwise directed by the service literature for the set). The other lead should be connected to the chassis of the receiver.

To arrive at the best tuning, the trimmers and padder are adjusted for a maximum reading on the output indicator. As shown in the diagram of Fig. 7-50, the trimmers are in parallel across the respective sections of the main tuning capacitor. The padder is in series with the coil and the main capacitor section of the oscillator tank. In transistor circuits, the capacitor might look somewhat different than what is illustrated here.

Not only must the carrier and oscillator tank circuits be tuned for maximum performance, but their resonant frequencies must differ by 455 kHz.

First, set the signal generator to 1500 kHz, leaving the audio modulation on. Rotate the main tuning capacitor of the re-

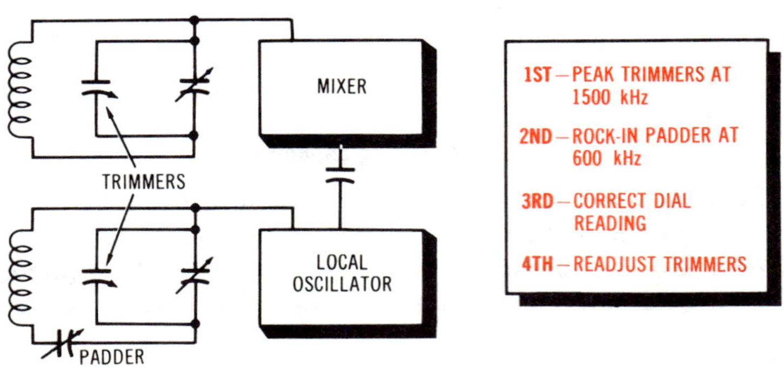

Fig. 7-51. Front-end tuning.

ceiver to the position of maximum output at the high end of the dial. Adjust both trimmers for highest output.

Now, set the generator and receiver dials to 600 kHz. **Rock-in** the padder tuning by making the same adjustment at several settings on either side of the 600-kHz dial reading. The setting at which the highest output reading is noted is the true 600-kHz position. Adjust the receiver dial pointer to indicate 600 kHz. Recheck the trimmers for maximum output again, first at 1500 kHz and, then, at 600 kHz. If no change is evident, tuning is completed.

TROUBLESHOOTING A RADIO RECEIVER

A signal generator is frequently used for troubleshooting a receiver. The signal generator can supply the proper signal through all of the stages except two—the power supply and local oscillator. The local oscillator generates its own frequency, and little test value is gained by passing another signal through it.

Two different methods are used to isolate the trouble in a receiver by means of a signal generator. One is called **signal substitution,** and the other, **signal tracing.**

Signal Substitution

In signal substitution, the output of an rf signal generator is applied to each stage in sequence. The faulty stage lies between the points at which the generator did and, then, did not pass a signal through the receiver.

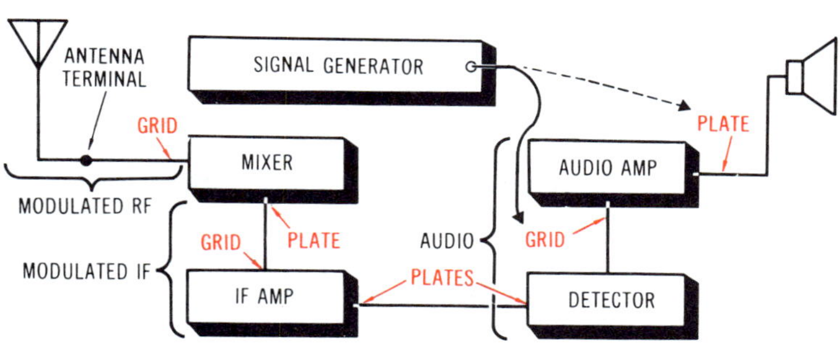

Fig. 7-52. Signal substitution.

The principle of signal substitution is simply to start at the output end of the receiver and work toward the front end, applying the appropriate signal at each stage (Fig. 7-52). Use a capacitor in series with the test lead to keep dc out of the generator.

The first check is made at the plate of the audio amplifier with an audio signal (only) from the generator. If a sound is heard in the speaker, the speaker and the circuitry between the speaker and audio amplifier plate are not the cause of the trouble. If there is no output from the speaker, the trouble is in this area. The trouble could be in the speaker, the output transformer, or in the several connections between the speaker and amplifier.

If the first check is good, apply the audio signal to the control grid of each audio stage. Audio from the speaker indicates that these stages are probably good.

Set the generator for a modulated if, and connect its output to the plates of the detector, the plate and grid of the if amplifiers, and the plate of the mixer. Apply modulated rf to the grid of the mixer and the antenna terminal. Somewhere in the sequence, no signal will reach the speaker. The trouble is between that point and the last good check point.

Signal Tracing

Signal tracing is almost the inverse of signal substitution. The signal generator is set up to provide a modulated rf sig-

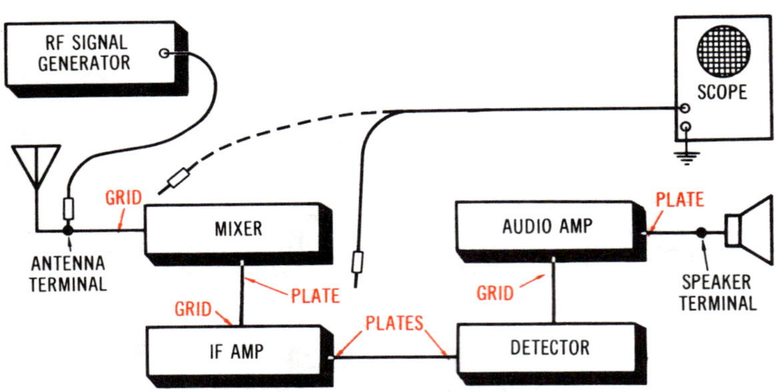

Fig. 7-53. Signal tracing.

nal within the range of the receiver; the test lead is connected to the antenna terminals. An indicating device—oscilloscope, vtvm, or solid-state vom, for example—is used to trace the signal from the front end to the output end of the receiver, making checks at the input and output of each stage. (See Fig. 7-53.)

An oscilloscope is best for signal tracing. You can use a vtvm or solid-state vom, however, to check for the presence of a signal. You must use an rf probe with the meter when testing at points where either rf or if signals are present.

The checks are made first at the input to the mixer, then at its output, and so on down the line from the input to the output to the input until no output appears at a check point. The trouble is located between this point and the previous point where an output was indicated. Use a vtvm or other type of multimeter in that area to find the faulty part.

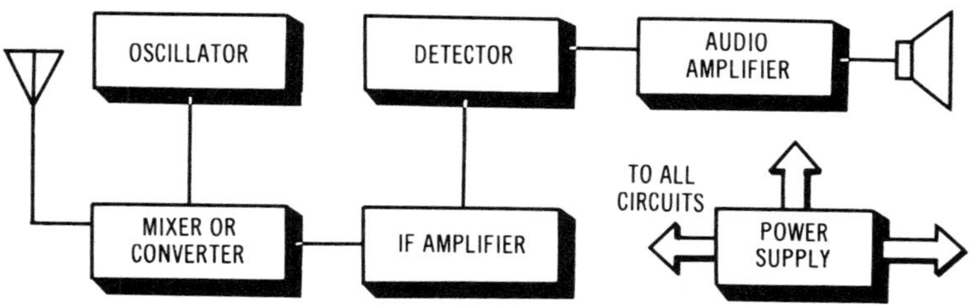

Fig. 7-54. Block diagram for Problem Q7-40.

The signal-tracing method can also be used for keeping a record of receiver sensitivity. At a time when the receiver is known to be in a satisfactory condition, connect the generator to the set, as indicated in the circuit for signal tracing (Fig.

Q7-36. What two signal-generator methods are used to isolate trouble in a radio receiver circuit?

Q7-37. Which method of receiver troubleshooting using a signal generator is better, signal substitution or tracing?

Q7-38. If you were using a vtvm as the indicating device in signal tracing, at what check point, in the receiver diagrammed in Fig. 7-54, would you change from an rf probe to the normal ac probe?

Your Answers Should Be:

A7-36. The two methods used to isolate a bad circuit are **signal substitution** and **signal tracing.**

A7-37. **Both methods are equally good.** The choice is merely a matter of personal preference.

A7-38. In the front-to-back signal-tracing method, the rf probe should be used last **at the plates of the detector.**

7-53). Set the generator and receiver gain to the minimum level that will cause a readable signal to pass through the set. Record the amplitude of the signal at each check point, using a vtvm or oscilloscope. Also, record the setting of the gain control and attenuator on the generator. If you wish, you can divide the output reading of each stage by its input reading to determine gain—the factor for sensitivity. These readings can be used at a later date for comparison purposes when a decrease in sensitivity is suspected. A loss of gain in a stage will indicate actual or impending trouble.

WHAT YOU HAVE LEARNED

1. An rf signal generator has much in common with a radio transmitter. They both have a means of generating a frequency, modulating it, and applying the output signal to a load. For this reason, a signal generator can take the place of a transmitter when you are checking the performance of a receiver.
2. Signal generators can be classified in terms of the frequencies generated by their main oscillators. An af (audio frequency) generator can develop a signal within the 20- to 20,000-hertz range. An rf (radio frequency) generator can have a range starting at about 20 kHz and reaching as high as several thousand megahertz. No single generator, however, can cover the entire range. Most rf generators have a means of modulating their carrier frequency.
3. There are several types of modulation. The two most used are amplitude modulation (am) and frequency modulation (fm).

4. There are many types of oscillators. Each type requires a means of amplification, a method of feedback, and some manner of frequency control.
5. The frequency at which an oscillator operates is controlled by the inductance and capacitance in the tank circuit. If either L or C is increased, the resonant frequency will decrease. Decrease either L or C, and the frequency will increase.
6. Oscillators in a resonant tank are damped (die out) because of the resistance (coil and conductors) in the circuit. Feedback from the amplifier plate, if properly applied, provides the periodic surge required to keep the oscillations going.
7. An af signal generator can be used with an oscilloscope to plot the frequency-response curve of an amplifier.
8. When the output of an af signal generator is fed into an overdriven amplifier, the output will be a fairly good square wave. Since a square wave is rich in harmonics (multiples of the fundamental frequency), the frequency response of an amplifier can be studied by observing the type of distortion produced in the square wave as it passes through the circuit.
9. Function generators provide any of a number of output waveforms, such as sine waves, sawtooth waves, triangular waves, and square waves. The waves can be varied in frequency and amplitude.
10. A typical rf generator contains an rf oscillator, an audio oscillator, a stage for mixing the two, an amplifier, a means of controlled attenuation, and an output meter. Such a generator provides a pure rf signal, an audio signal, or a modulated rf signal.
11. In a superhet receiver, modulated rf and the oscillator frequency are applied to the mixer stage. They are heterodyned to produce a sum and a difference frequency. The difference frequency, still containing the modulation envelope, is referred to as the if frequency and is amplified in the next stage. The detector extracts the audio signal from the if carrier and feeds it to the audio amplifier and, then, to the speaker and, also, provides avc. Some superhets have an additional stage, called an af amplifier, ahead of the mixer. A converter

serves the combined functions of the mixer and oscillator stages.

12. Alignment of a superhet receiver is a relatively simple task if the technician understands how his signal generator and receiver operate. Specific information for the generator is contained in the operating manual and, for the receiver, in a service folder.
13. The steps for receiver alignment are:
    (A) Tune the if transformers.
    (B) Tune the front end.
14. An rf signal generator can also be used for isolating trouble in a receiver to one of its stages. There are two methods used—signal substitution and signal tracing.

8

Troubleshooting Electronic Equipment

what you will learn

The primary task of any electronic technician is to troubleshoot the equipment. Since most technicians have difficulty in acquiring a reliable method, this chapter will explain one that is used by most technicians. Some technicians call it systematic troubleshooting; others call it common-sense troubleshooting. The title that seems to embody both names, and the one that will be used in this chapter, is logical troubleshooting. There are many methods other than the one that will be described; however, by comparison these methods have been found to be ineffective and time consuming.

THE NEED FOR TROUBLESHOOTING

As you have already determined, troubleshooting is the process of locating the fault that causes a piece of equipment to operate at less than desired or designed performance.

Any equipment operating at less than the best performance requires the services of a troubleshooter. A stereo set that is garbling its highs or lows, even though it has a good rated frequency response, has an electronic fault that needs repair. A home radio that begins to pick up two stations at once contains a defect. A tv set that has poor contrast between blacks and whites, or the wrong colors, also needs repair. The remedy may be no more than the proper adjustment of one or two controls, but the trouble will remain until the appropriate adjustment is located and made.

The need for troubleshooting (locating the cause of faulty performance) exists whenever the equipment fails to meet the rated performance as set forth by the manufacturer (Fig. 8-1).

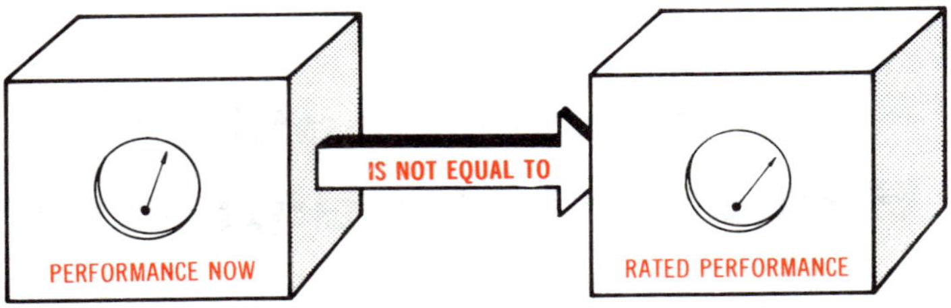

Fig. 8-1. Equipment does not meet rated performance.

TROUBLESHOOTING PREREQUISITES

Good troubleshooting is not a talent with which a person is born. It is a skill that can be acquired by anyone with a suitable electronics background. You can become a good troubleshooter if you have:

1. Sufficient electronic knowledge to learn how a piece of equipment works.
2. Suitable skill in reading and interpreting data contained in the technical manual or service folder.
3. Suitable skill in operating test equipment and interpreting test readings.
4. The ability to troubleshoot in a logical manner.

Electronic Knowledge

If you have carefully studied the preceding volumes in this series (or have an equivalent knowledge) and have been able to apply these electronic principles, you can learn how electronic equipment works. You will encounter many circuits and pieces of equipment that are not familiar as you gain experience, but gaining an understanding of how they work is merely a process of applying what you have already learned.

What is this foundation that can be applied to all electronic devices? The answer is the set of principles that you learned about dc and ac electricity and which you have been applying in this volume.

The illustration in Fig. 8-2 shows that **all electronic equipment** is made up of or based on **selected electronic circuits** which, in turn, operate in accordance with the fundamental principles of voltage and current, and the characteristics of

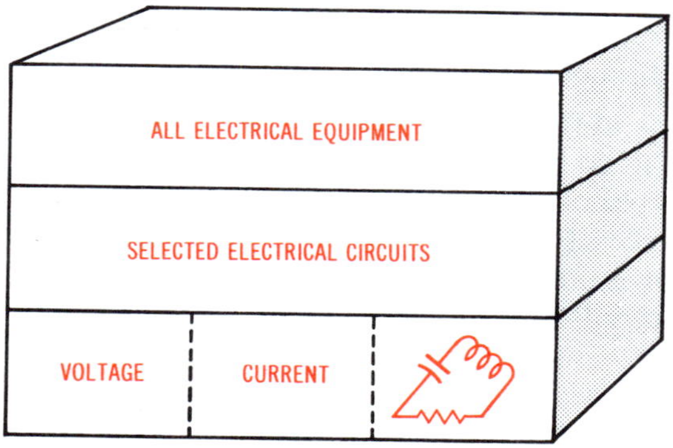

Fig. 8-2. Electronic building blocks.

inductance, capacitance, and resistance. If you reduce any electronic equipment to the bare essentials, you will find that the equipment operates the way it does because of the circuit arrangement of L, C, and R and their effect on current and voltage. If you have the foundation for understanding how electronic equipment operates, you need only experience and more study if you wish to become skilled.

Reading and Interpreting Electronic Data

Most electronic devices have operating or servicing manuals, often called technical manuals or instruction books. They contain text, diagrams, and other data required for troubleshooting. Equipment used in the home, such as radios, television receivers, and audio equipment, usually has service folders that contain similar information. These service folders can usually be procured from any local electronic-parts supply house.

Q8-1. What is troubleshooting?

Q8-2. What are the requirements for a good electronics troubleshooter?

Your Answers Should Be:

A8-1. Troubleshooting is the process of **locating** the causes of malfunctions in an electronic circuit.

A8-2. The requirements for a good troubleshooter are a basic **knowledge of electronics, an ability to read and interpret data,** the **ability to operate various test equipment,** and a **logical testing method.**

Reading Technical Data

Will you be able to read these manuals and folders? Yes, if you were able to understand the information presented thus far in this volume. The portions of schematic and block diagrams, the circuit descriptions, and the kind of test data that you have encountered are all representative of the information that you will find in the manuals and folders.

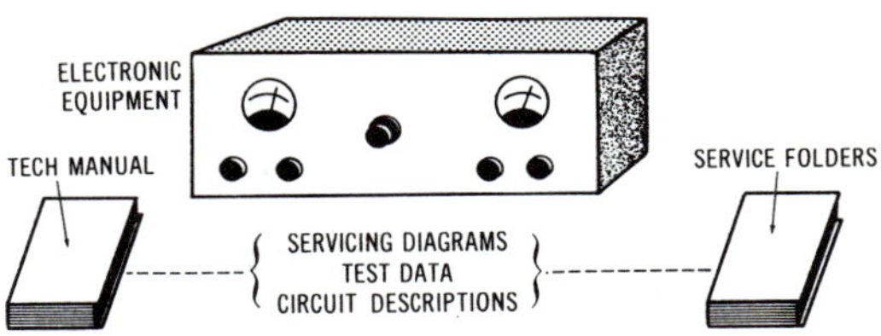

Fig. 8-3. Servicing diagrams and other technical data.

Test Equipment

You have studied the basic types of test equipment. All other types of test equipment are more or less complex adaptations of those included in this volume. Like any electronic equipment, their operation is founded on a basic set of principles. Therefore, you have the capability of learning how to operate and use them. Again, experience is the instruction needed to gain greater skill.

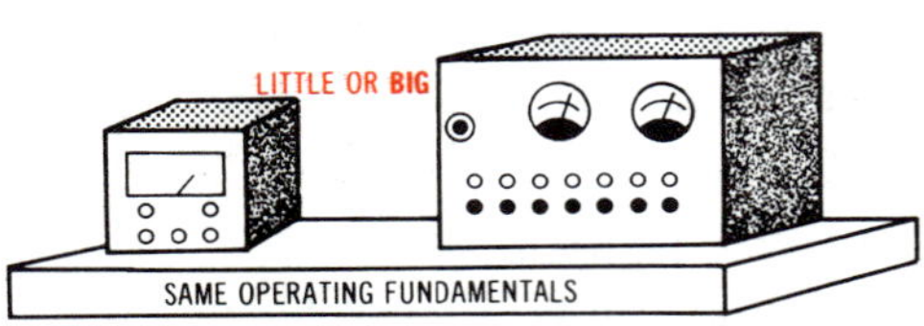

Fig. 8-4. All test equipment has a basic set of principles.

LOGICAL TROUBLESHOOTING

Logical troubleshooting is a systematic, common-sense method of isolating the fault in a malfunctioning piece of equipment. It does not employ the time-wasting or ineffective procedures of trial-and-error methods. The logical troubleshooter uses his knowledge of electronic principles, his ability

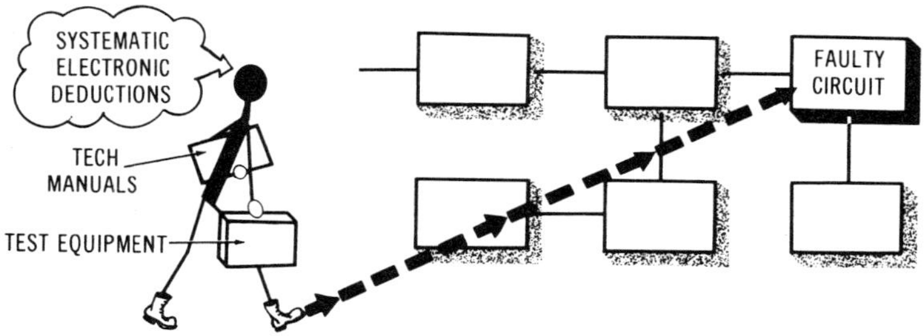

Fig. 8-5. Logical troubleshooting.

to extract data from a technical publication, and his skill in using test equipment.

Logical troubleshooting is a time-proven procedure used by all experienced technicians. Most of them have applied the procedure so often that they no longer pay attention to its fine points. Through habit and years of experience, they may have forgotten the specific details.

Probably no two technicians would explain the procedure alike. However, all would agree that logical troubleshooting consists of a series of sequential steps. Each step is based on valid electronic deductions that systematically narrow down the trouble to increasingly smaller areas in the equipment and, finally, to the faulty part, wire, or connection. Some technicians

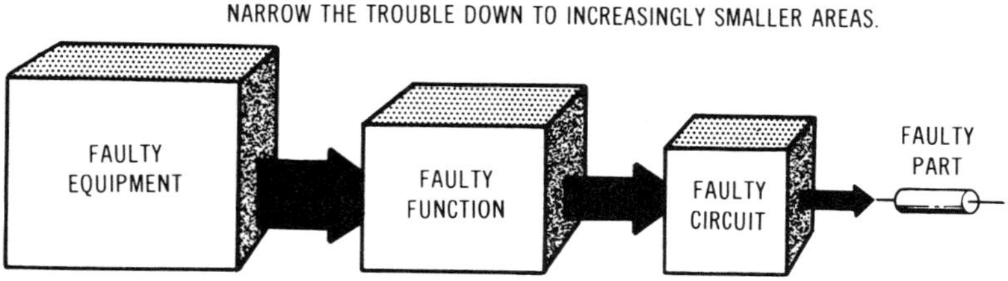

Fig. 8-6. Use a series of sequential steps.

might list the procedure in two or three steps; others would count a dozen or more. Regardless, the principle is the same.

Five steps are listed below as the most reliable method of learning and applying this procedure. They can be applied to any equipment, regardless of size. The steps in the proper order are:

STEP 1. Search for all trouble symptoms.
STEP 2. Trace out all probable faulty functions.
STEP 3. Expose the single faulty function.
STEP 4. Pick out the faulty circuit.
STEP 5. Seek out and verify the cause of the trouble.

Note that the first letter in each step spells STEPS (Fig. 8-7). This fact will help you to remember logical troubleshooting.

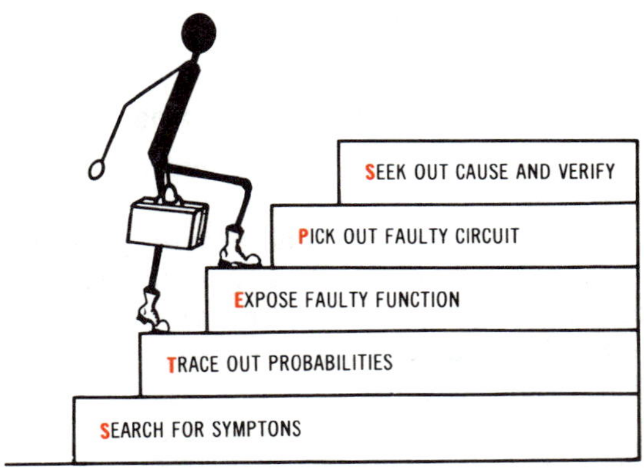

Fig. 8-7. The five steps of logical troubleshooting.

STEP 1. SEARCH FOR ALL TROUBLE SYMPTOMS

A trouble symptom is an outward indication that a piece of equipment is not working properly. In dead equipment, the indication is fairly obvious. A hum in a radio receiver, a distorted picture on a tv set, or harsh flat notes from a hi-fi set are also obvious, and make further use of the equipment undesirable. Then, there are the less obvious indications as the performance of the equipment slowly worsens over a period of time. These are tolerated until the output becomes obviously distorted or blanked out.

Symptom Indicators

Audible and/or visual outputs of an item of equipment are symptom indicators which, by the use of the front-panel controls, can help you to pinpoint the source of trouble.

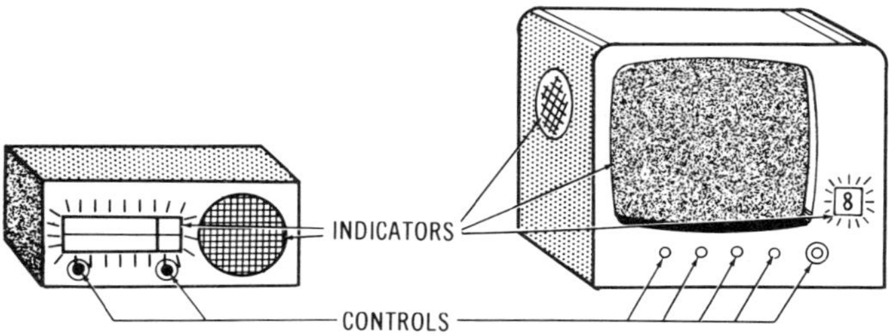

Fig. 8-8. Indicators and controls.

Many radio receivers (Fig. 8-8) have two output indicators, a speaker and a light (usually illuminating the dial) ; the receiver also has at least two controls. The change from desired performance can be registered in many ways—hums, squeals, squawks, low volume, two stations instead of one, or no sound at all. The light is either on or off.

The controls can be used to obtain more information about the symptom. How does the audio change, if at all, when the volume control is rotated from one extreme to the other (Fig. 8-9) ? Does the hum or other noise become louder, or does it remain the same? If there is no undesirable noise, will the control smoothly increase the volume of the station program?

Fig. 8-9. Listen to the sound output.

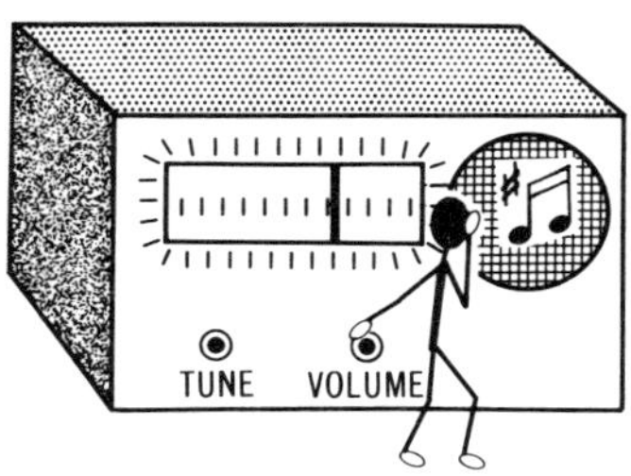

Q8-3. If the dial light is out, what are the possible causes of the trouble symptom?

Q8-4. If you have determined that a radio is receiving only two stations instead of many, would this fact be a useful trouble symptom?

Your Answers Should Be:

A8-3. If the dial light is out, the trouble could be **no power to the radio, a burned-out lamp, a faulty switch,** or **faulty conductors, connections, or parts** within the set.

A8-4. **Yes.** If this information is not determined before proceeding further into the troubleshooting procedure, much time could be spent in checking unnecessary areas of the radio.

Obtain as much symptom information as you can during Step 1. Learning as much as you can about the trouble symptoms is the only effective way to begin a search for the cause and its source.

A television receiver has an additional output and a greater number of front-panel controls that can be used in searching for trouble symptoms (Fig. 8-10). It has a speaker, a dial light,

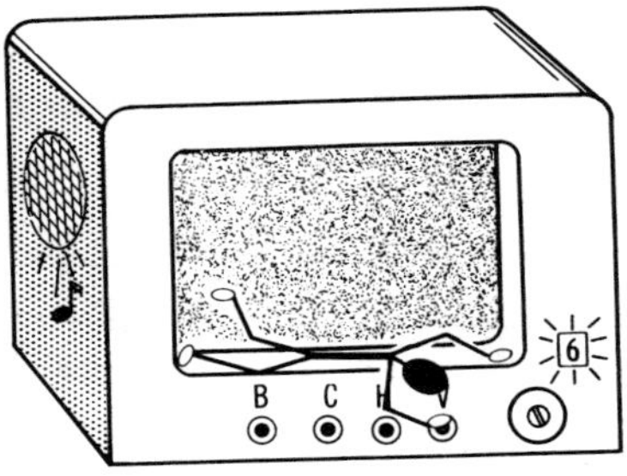

Fig. 8-10. A tv receiver has many controls.

and a visual output indicator to detect trouble symptoms, and several controls that can be adjusted to observe additional symptoms or changes in output. Another advantage in first looking for all symptoms is that proper adjustment of a control will quite frequently eliminate the trouble. Ragged, slanted lines on the tv screen might be corrected by adjustment of the horizontal control. Distortion in height of the picture (large heads and short legs, for example) might be corrected by adjustment of the vertical-linearity control. If these adjustments correct the fault and there are no other symptoms, the troubleshooting job is completed.

The more complex equipment used in commercial, industrial, and service installations provides additional means for tuning and setting-up operations.

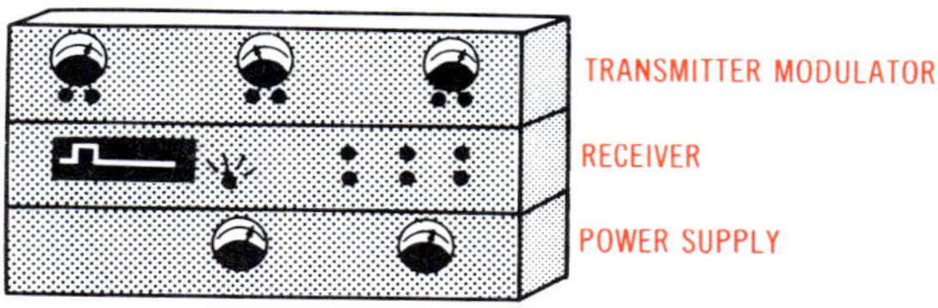

Fig. 8-11. Commercial equipment is more complex.

STEP 2. TRACE OUT ALL PROBABLE FAULTY FUNCTIONS

As applied to electronic equipment, a **function** is the purpose of the equipment, group of circuits, or circuit (Fig. 8-12). In the narrowing-down feature of logical troubleshooting, the idea is to pick out a few of the several functional circuit groups in which the trouble most probably lies. When this is accomplished, the search is narrowed down to a smaller area.

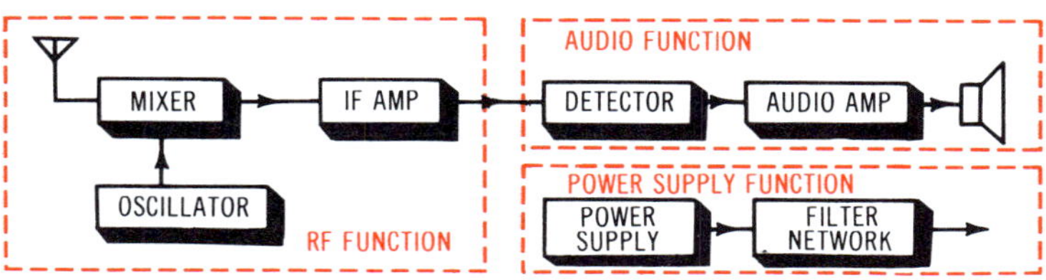

Fig. 8-12. Partitioning the equipment into its functions.

In smaller equipment such as the radio receiver illustrated in Fig. 8-12, the number of circuit-group functions may be limited. Further limitations are imposed by the receiver, which has only two controls—tuning and volume. However, the antenna, mixer, oscillator, and if amplifiers of the set shown can be grouped within a radio-frequency function. The combination of detector, audio amplifier, and speaker is designated as an audio function; the power supply, its filter network, and power cord become the power-supply function (Fig. 8-13).

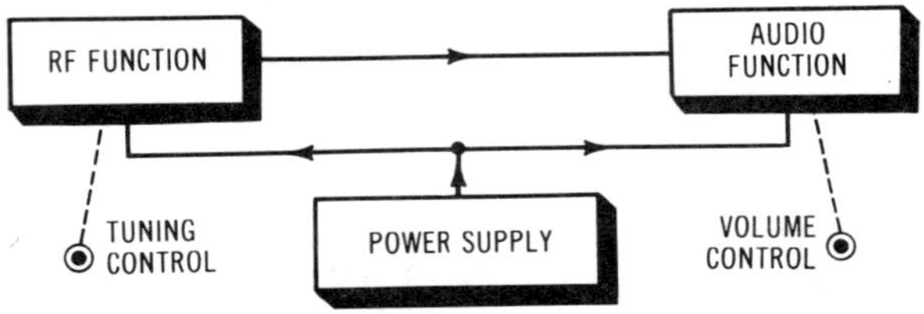

Fig. 8-13. Functional circuit group.

Isolating Faulty Radio Functions

The tuning control of the receiver is connected to the inputs of the mixer and oscillator, and the volume control is connected in the input circuit of the first audio amplifier stage. Information obtained from adjustment of these controls can, therefore, be associated with the respective functional groups. The purpose of the second step is to trace out and identify the functions whose symptoms indicate a malfunction. The following is an example of the second step. The original symptom in the radio receiver is weak output. Adjusting the volume control makes little or no difference. The tuning control shows a small but significant difference between loud and weak stations. The dial lamp is on. In which function(s) is the probable location of the trouble? The most probable location is the audio function. The power-supply function is a possibility, but the rf function can almost be eliminated. If you were to list all three functions as probables, based on your technical knowledge of how the receiver works, your answer could be just as correct as the one given. Trace out all the probable faulty functions. Place them in the most logical order.

Try another problem. During Step 1, no stations are heard, regardless of where the tuning control is set. The dial lamp is on. Rotation of the volume control causes an increased crackling and rushing noise in the speaker. In which function(s) would you expect to find the trouble?

The most probable location of the trouble, in this case, is the rf function. The noise heard in the speaker is the normal noise generated by the vacuum tubes or transistors in the radio. This noise indicates that the tubes or transistors are getting voltage from the power supply and that the audio stages are performing their amplifying function.

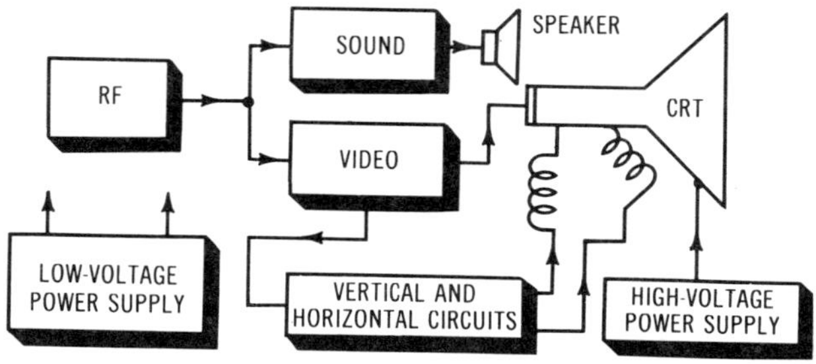

Fig. 8-14. Functional groups in a tv receiver.

Television Functions

A television receiver can be broken down into a large number of sharply defined circuit-group functions. The 30 or 40 circuits in one type of television receiver can be visualized in functional circuit groups as shown in Fig. 8-14.

Suppose that during Step 1, you obtained the following symptoms. Audio appears to be good, but the picture covers only half of the screen vertically. Width appears to be proper. Adjustment of the vertical control makes no apparent change in height, but it does cause the picture to roll.

Since audio (sound) and image (picture) appear to be good, the sound and video functions are eliminated. If these are working properly, the rf function must be operating properly. If the high voltage were low or absent, there would be no picture on the screen. The low voltage must be good since the picture, its width, and the sound are good. Logical reasoning indicates that the trouble must be in the vertical and horizontal circuits.

As a result of reaching only the second step in logical troubleshooting, you have limited the trouble to a half-dozen circuits out of a possible thirty or forty. This is much better than checking them all. In addition, the logical deductions you have made have given you a good idea as to the type of trouble you are looking for. Evidently, the output voltage of the vertical-deflection signal is not large enough to swing the electron beam over the full height of the screen.

Q8-5. Which function in the diagram of Fig. 8-14 is the most probable location of trouble if the picture is good but there is no audio output?

> **Your Answer Should Be:**
>
> A8-5. **Sound** (including the speaker) is the only probable malfunction. A good picture indicates that the other functions must be working properly. The rf and low-voltage power-supply functions feed both the sound and picture sections; since the picture is proper, both of these functions must be working.

Review of Steps 1 and 2

During the first two steps of the troubleshooting procedure, analysis is confined to information obtained from outside the equipment (Fig. 8-15). After obtaining all the information

Fig. 8-15. You have taken the first two steps.

you can about the original trouble symptom(s) by manipulation of front-panel controls and observation of output indicators (Step 1), you proceed to the second step and trace out all probable faulty functions. In Step 2, you use symptom information to make logical technical deductions and identify the functional areas of equipment that may contain the trouble. Up to this point, you have neither entered the equipment nor used any external testing devices.

While making technical deductions during Step 2, you may find it desirable to obtain additional symptom information. Returning to the procedures of Step 1 will not violate any rules. You may often find it necessary to return to a previous step or steps for re-evaluation purposes. Until you become experienced in troubleshooting, write out the data obtained or conclusions reached during each step. You will find that this procedure reduces the necessity of returning to a previous step for verification.

STEP 3. EXPOSE THE SINGLE FAULTY FUNCTION

In Step 3, you can use test equipment to determine which one of the probable faulty functions contains the trouble.

Radio Application

Refer to one of the radio receiver examples used in the preceding step as the first example. No stations were heard at any frequency. The lighted dial lamp indicated that power was applied. An increase in receiver background noise as the volume control was adjusted indicated that the audio function was good. It was decided that the radio-frequency function was suspected to be defective.

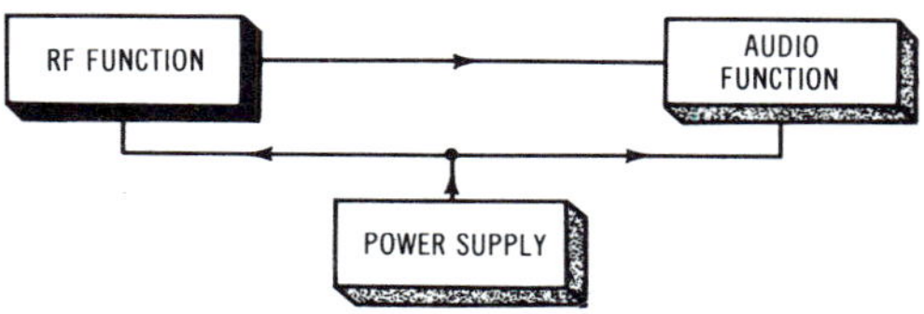

Fig. 8-16. You suspect the rf function is defective.

The only purpose of Step 3 is to locate the single function that is causing the equipment to operate improperly. In the example of Fig. 8-16, either a scope or a good electronic volt-

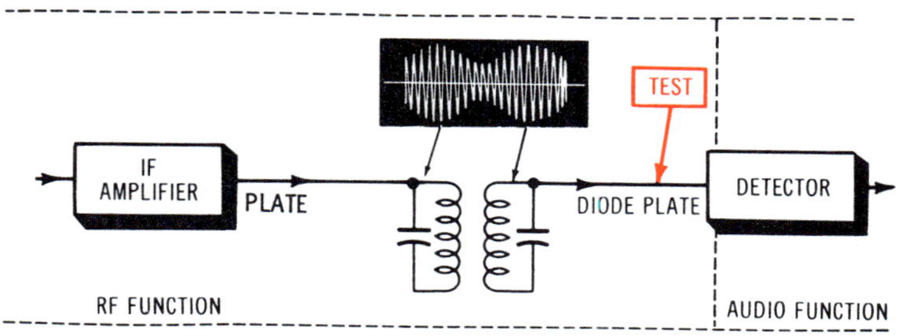

Fig. 8-17. Testing for presence of a signal.

meter can be used. From the schematic diagram of the receiver, locate the pin number of the upper diode plate of the detector (Fig. 8-17). Connect the oscilloscope to the proper socket terminal and rotate the tuning control. If no audio-modulated sig-

Q8-6. What would you look for with a vtvm or a solid-state vom?

Your Answer Should Be:

A8-6. You would look for **an ac signal.** If this signal is present, the voltage would not be very large and the meter reading would show a steady average of the changing amplitude. For this reason, an oscilloscope would give better confirmation of a signal if one were present.

nal is noted at several station settings, the rf function is not operating properly.

If a good, but weaker-than-normal, signal is obtained, deductions made during Step 2 are erroneous. However, the effort made thus far in Step 3 is not wasted. You have added more data to your store of information. You can now go back to the procedures of Step 2 and the functional block diagram better equipped to select the probable fault(s). Having recorded a weak output for the rf function, you conclude that the weak signal should have been passed through the audio function if it were good. When a recheck is made of speaker output with the volume control at maximum, background noise this time seems weak. Since both functions are weak, you suspect the power supply of faulty performance (Fig. 8-18).

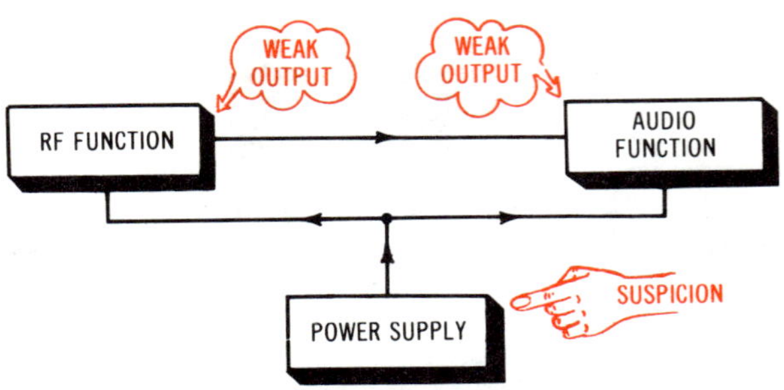

Fig. 8-18. The power supply is suspected.

The new Step-2 conclusion places the probable location of the trouble in the power-supply function. A Step-3 check of the schematic for the receiver indicates that there should be a pulsating dc output of 90 V. With a vtvm or a multimeter, the dc reading shows less than half this value. The faulty function has been confirmed.

Tv Receiver Application

The results of the first and second steps for a tv receiver could be the following:

STEP 1. **Symptoms**—Good audio; good picture image but it covers only half the height of the screen. There is no change while adjusting the vertical control.

STEP 2. **Deductions**—The trouble is probably in the vertical and horizontal functions (Fig. 8-19).

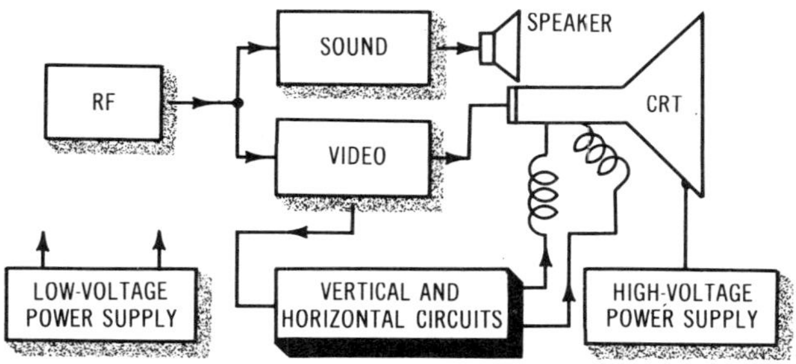

Fig. 8-19. The vertical and horizontal functions are suspected.

The schematic diagram included in the technical manual or the servicing folder for the receiver should be used in locating the output test point of the probable faulty function. You will find that a schematic diagram will be your most valuable single item for troubleshooting.

The oscilloscope is the best piece of equipment to use for obtaining readings at a suspected trouble point, since you can observe the shape of a waveform as well as measure the amplitude (voltage, in this case). To measure voltage, the oscilloscope screen with its graticule must first be calibrated from a known voltage source. Multimeter or vtvm ac voltage readings are difficult to use for confirming whether the output is good or bad. The tv receiver schematics usually show waveform outputs with peak-to-peak (p-p) voltage values. These are difficult to convert to meter readings.

Q8-7. How would you locate the correct test point to verify that the trouble is in a certain function?

Q8-8. What test equipment would you use to check the waveform at the test point?

Your Answers Should Be:

A8-7. Use the **schematic or block diagram** that comes with the equipment to locate individual test points.

A8-8. The **oscilloscope** is the best piece of equipment to use when checking waveforms.

A low reading would substantiate the Step-2 deduction that the vertical and horizontal circuits probably contained the fault. However, care should be taken in making this decision. The reading should be substantially lower than that shown in the service data—about half as much in this case. Since there is a variation in part values among pieces of equipment, test values on a diagram are only representative of those found in most equipment of the same model. However, the equipment readings should be within a few percent of the values specified. If the output of the vertical section reads low in this example, Step 3 would be successfully concluded. If the reading is very close to normal, your conclusion must be that the vertical and horizontal functions are probably not at fault. If the oscilloscope test produced these results, what should you do next? Revert to Step-2 procedures, trace out all probable faulty functions, and then apply the new information you have learned.

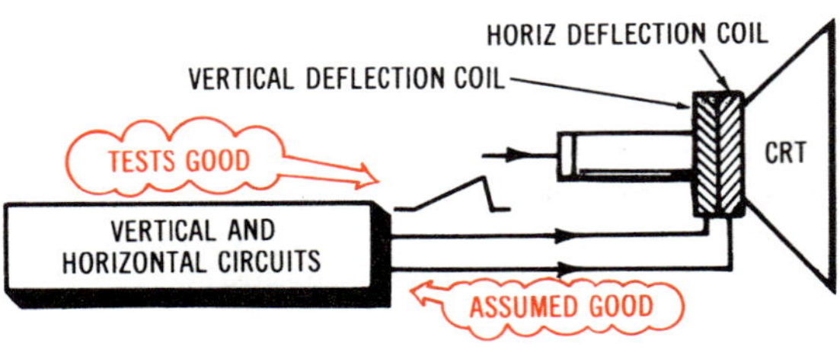

Fig. 8-20. Checking the vertical and horizontal circuits.

In re-evaluating your symptom information on the Step-2 level (Fig. 8-20), you find:

1. Good sound and picture image, therefore, power-supply voltages must be correct.
2. Horizontal width of the picture on the screen seems proper, so that portion of the circuit can be assumed to be good.

3. The verifying test showed that the vertical output was operating as it should. Frequency, shape, and amplitude of the sawtooth output appeared good.

Since deductions and tests show that all other functions are good, the possibility is very strong that the fault is in the vertical-deflection coil, since this is the only remaining part that has any control over the height of the picture.

Now, apply Step 3 to a more difficult tv malfunction. This is what you might write down for the first two steps:

STEP 1. **Symptoms**—Sound is good, but weaker than normal. The screen is blank; there is no picture or raster (horizontal lines on the screen when station is not on the air). The adjustment of contrast, brightness, vertical, horizontal, and fine-tuning controls makes no change. Moving the channel-select knob to other stations has the same results.

STEP 2. **Deductions**—Sound and rf functions are probably good. The low-voltage power supply might be good, since the sound circuits are operating. However, the power supply might be providing just enough voltage for sound and rf, but the output is too low for one or more of the other functions. All the other functions—video, crt, high-voltage power supply, and horizontal and vertical circuits—are probable causes of trouble.

In the functional block diagram of the tv receiver shown in Fig. 8-21, the suspected functions are marked with **PF** (probably faulty), and the unsuspected, with **PG** (probably good).

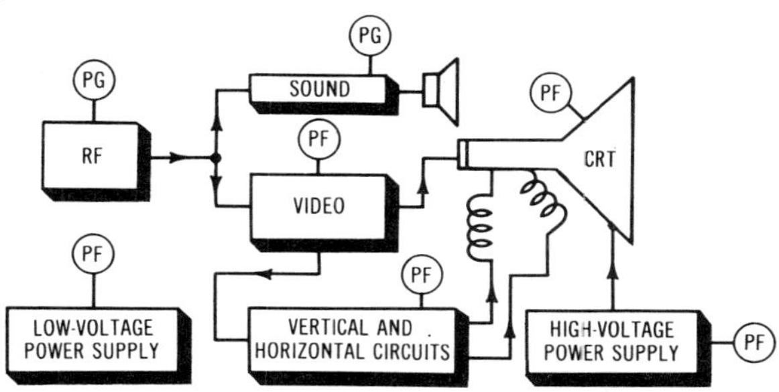

Fig. 8-21. Functional block diagram of a tv receiver.

Five functions are suspected as probably being faulty. In which order should they be tested to arrive at a Step-3 conclusion (exposure of the single faulty function)? Three rules should be applied in answering this question. First, make only those tests that are safe to make. Second, make the tests in the order of least difficulty. A test that requires dismantling a section of the equipment is an example of a difficult test. Third, test those functions first that will eliminate one or more of the other functions being considered as probably faulty. Those that are equal in terms of these rules become a matter of personal choice in the testing sequence. A good sequence of function tests is the following:

1. Vertical and horizontal circuit function—This is selected first because a raster would be present on the screen if it were operating properly, whether the video function is sending sync signals to it or not. Under the conditions of a completely blank screen, there is neither a vertical nor a horizontal output, if this is the faulty function. If the function checks to be good, with an oscilloscope test of the two outputs, the low-voltage supply is considered good. Sufficient voltage is being applied to operate the sync function.
2. Cathode-ray tube—The crt should be checked before the high-voltage power supply. By looking down into the base of the tube you can determine whether the heater is working. If it is, there will be a bright glow. Also, check for gas. This is determined by a bluish glow within the neck of the crt. A small blue cloud near the base, although not desirable, will have little effect on the beam and does not explain a blank screen.
3. Video function—Although the output of the video function could not be responsible for the missing raster, the function is worth testing. Video output to the crt is checked for proper values of image and blanking pulse, and the output of the sync function is measured for sync pulses.
4. High-voltage power supply—If all the preceding functions have been tested and rated as probably good, the high-voltage power supply could be exposed as the single faulty function by default. A review of the symptoms and

test results makes this a logical deduction. If high voltage is missing from the crt, the electron beam will not reach the screen. As a result, neither a raster nor a picture will appear.

Review of Steps 1, 2, and 3

The STEPS procedure has now been three-fifths completed. The approach has quickly narrowed the trouble to a single function among several by making logical technical deductions on the basis of accumulated data.

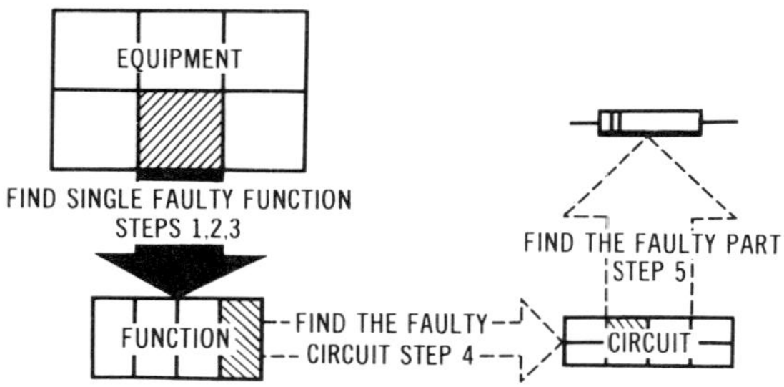

Fig. 8-22. The STEPS method.

STEP 4. PICK OUT THE FAULTY CIRCUIT

The narrowing-down process continues in the fourth step by working toward the faulty circuit within a functional group. The procedure is carried out by making technical deductions from accumulated symptom and test data. These deductions result from studying the servicing block diagram and, then, closing in on the malfunctioning circuit.

The Servicing Block Diagram

This is a diagram that you have used many times in this volume. It consists of individual blocks representing each circuit within the functional group. The blocks are interconnected to show the direction of signal flow, and the input and output test points are indicated. Some servicing block diagrams include waveform data at significant points within the diagram.

Q8-9. Testing which tv receiver function would violate the rule: "Make only those tests that are safe."?

> **Your Answer Should Be:**
>
> A8-9. The **high-voltage power supply** would be unsafe to test. The output could be as high as 30,000 V.

Quite often, the equipment you will be troubleshooting may not have a servicing block diagram. The schematic diagram can be used instead. However, until you become accustomed to visualizing the schematic of individual circuits as a simple functional block without the distracting influence of its individual parts, you should draw your own block diagram. The complete servicing block diagram for a radio receiver is shown in Fig. 8-23.

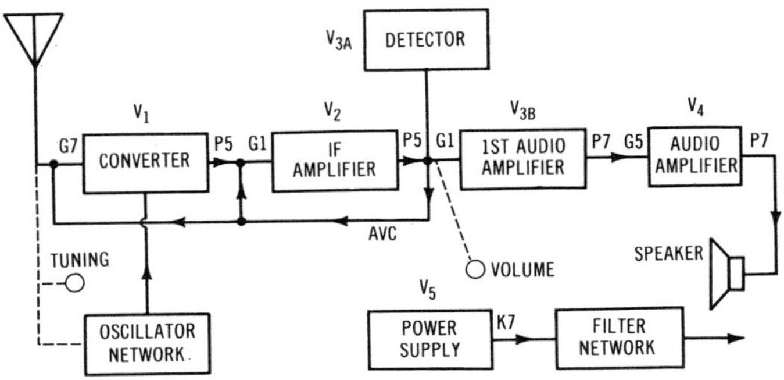

Fig. 8-23. A servicing block diagram for a radio receiver.

With waveforms shown between stages and with input and output test points identified, a servicing block diagram can be used for isolating a faulty circuit. In the diagram of Fig. 8-23, tubes V_1, V_2, and V_{3A} are included in the rf function; V_{3B} and V_4 are in the audio function. V_5 is the power-supply tube.

Closing-in Procedure

When picking out the faulty circuit in Step 4, it is neither desirable nor necessary to check the inputs and outputs of each circuit contained in the faulty function. Some functions may have two or three stages, and others, a dozen or more. Finding the faulty circuit with a minimum number of tests is accomplished by using a **closing-in,** or **bracketing,** process.

When working from a servicing block diagram that contains the faulty function identified in Step 3, enclosing indicators are placed at the inputs and outputs of the functional group. These can be pencil marks as shown in Fig. 8-24, or they can be small weights to eliminate damage from repeated erasures. You can even depend on your memory for locating and recalling the enclosing marks. A **G** (good) mark at the input(s) indicates that this point has been tested and found to be satisfactory. A **B** (bad) mark at the output(s) indicates a test has revealed the output waveform is improper or nonexistent.

Linear Circuits

The diagram of Fig. 8-24 shows circuits following each other in a line. Such an arrangement is known as a **linear signal**

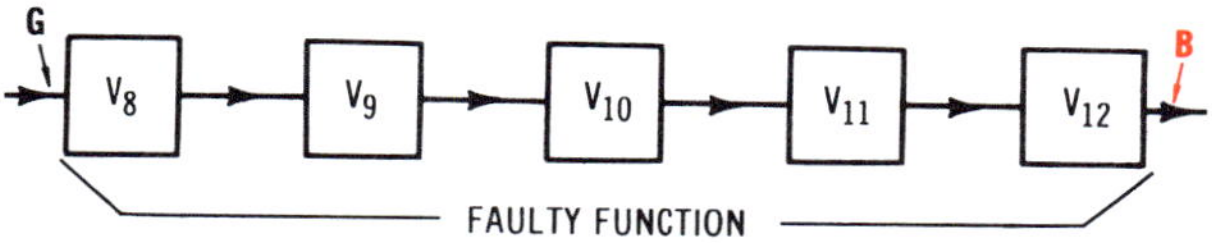

Fig. 8-24. Using indicators in a linear signal path.

path. Marks on the diagram show a good input and a bad output. The concept of Step 4 is to isolate the one faulty circuit among the five with the fewest tests.

To minimize the number of circuit tests required, the first check with the oscilloscope is made at either the input or output of V_{10}. Block V_{10} is the middle tube in the group; a good or a bad indication eliminates the necessity of checking about half of the circuits. It is usually acceptable to check the output of block V_9 or the input of V_{11}, since the waveforms at these points are essentially the same as the input or output of block V_{10}, respectively. This procedure of dividing a linear string of circuits for testing purposes is known as the **half-split method.** If the test is made at the output of block V_{10} and reveals an improper or nonexistent waveform, the bad indicator should be moved to that point.

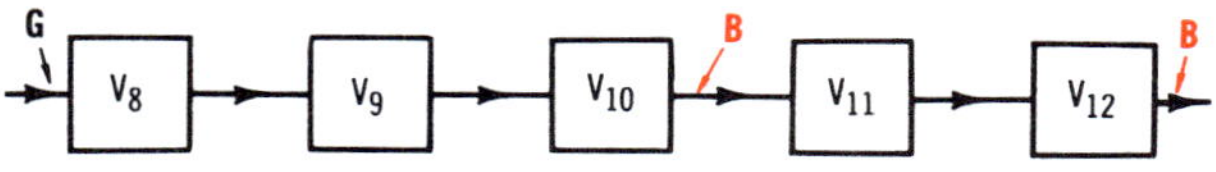

Fig. 8-25. The half-split method of testing.

Q8-10. What is the concept of Step 4?

Your Answer Should Be:

A8-10. The concept of Step 4 is to **isolate the one faulty circuit** among the many in the suspected function.

The faulty circuit is now located between the input of block V_8 and the output of block V_{10} (Fig. 8-26). If the scope test is properly made, tubes V_{11} and V_{12} are considered good. If the test reveals a proper waveform, the **good** mark is moved as shown below. With the **good** mark at the output of block V_{10}, circuits of tubes V_8 through V_{10} are no longer suspects; the

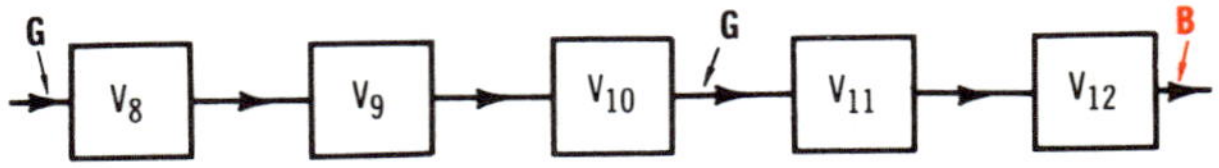

Fig. 8-26. Pinpointing the faulty circuit.

faulty circuit is, thus, block V_{11} or V_{12}. By moving either the **good** or **bad** mark, depending on the result of the test, the faulty circuit is restricted to a smaller enclosure. By half-splitting again between the new **G** and **B** marks, the enclosure is made even smaller. In the second example, a test at the output of block V_{11} identifies the faulty circuit. Depending on the results of the test, **G** or **B** is moved to that point, and either block V_{11} or V_{12} is pinpointed as the faulty circuit.

Take a dead receiver as a practical example. The diagram is shown in Fig. 8-27. The only symptoms obtained in Step 1 are no sound output or electrical power reaching the power supply. In Step 2, all functions are listed as probables. Step 3

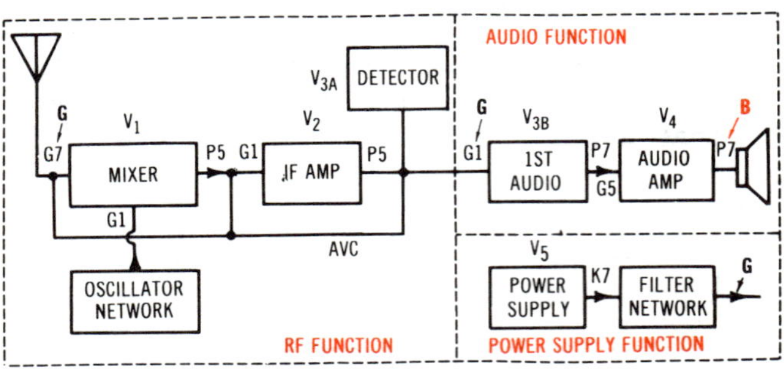

Fig. 8-27. Diagram for the "dead receiver" example.

reveals the following information. Good test indications are made at the input of block V_1, the input of block V_{3B}, and the output of the filter network in the power supply. A bad test is identified at the speaker input. The audio function is suspected of being faulty. The first test of Step 4 is made at the grid (pin 5) of audio amp V_4 or at the plate (pin 7) of V_{3B}.

Convergent Circuits

There are circuit combinations other than linear. One of these is called a **convergent** circuit. As the name implies, a convergent circuit is one in which the outputs of two or more circuits converge (join) to feed a single circuit (Fig. 8-28).

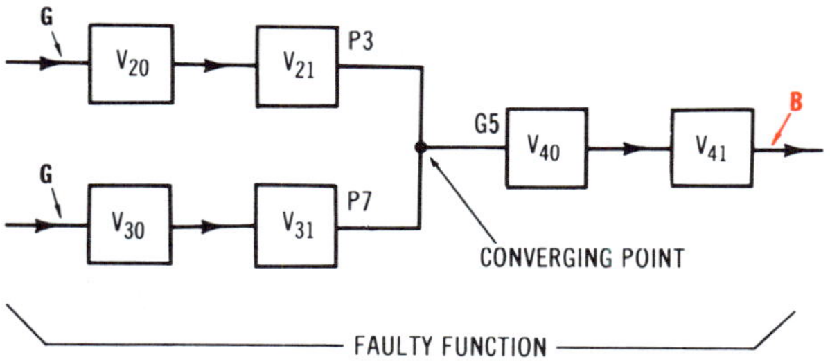

Fig. 8-28. Convergent circuit.

The diagram of Fig. 8-28 shows the test results of Step 3. Inputs to both channels of the function are good, but the single output is bad. The decision of where to make the first Step-4 test depends on the nature of the bad output.

First, assume that there is no output signal of any kind. After checking the function of block V_{40}, it is learned that this circuit does not operate unless the outputs of both channels are received. This is called a **gating** circuit. To minimize the number of tests, where should the first check be made?

The first test should be made at the converging point. A waveform reading at the input of block V_{40} will reveal the nature of the outputs from blocks V_{21} and V_{31}. If both waveforms were there and of the proper shape, the **G** could be moved to the converging point, thus limiting the remainder of the search to blocks V_{40} and V_{41}.

Q8-11. What is a convergent circuit?

Q8-12. What is meant by gating?

Your Answers Should Be:

A8-11. A convergent circuit is one in which the **outputs of two circuits join at a point** to supply a signal to one subsequent stage.

A8-12. Gating refers to certain circuits that have **one output but two or more inputs.**

It is not probable that both output signals are missing or improper. Therefore, the check at the converging point, if not good, identifies which channel is bad. Mark **B** is then moved to that output, reducing the number of faulty circuits to one. In cases where the convergent circuit passes either output signal as long as it appears on the grid, the approach to testing is a little different. Note the output waveforms in Fig. 8-29.

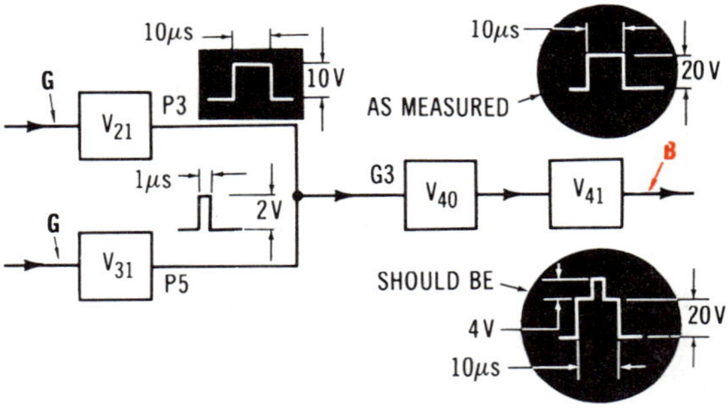

Fig. 8-29. A second test method.

The first test should be made at the converging point (G-3 of block V_{40}). A comparison of the two output waveforms—the measured waveform and the correct waveform—indicates that the 1-microsecond pulse is missing. A proper deduction shows that the small pulse does not leave block V_{31}, therefore, it is correct to make a test at the output of V_{31}.

If the preceding deduction is verified, mark **B** is moved to the output of block V_{31}. A single circuit is enclosed, and Step 4 is satisfactorily completed. But, if the test is good, the **G** at the input to block V_{31} is moved to the converging point, leaving blocks V_{40} and V_{41} as probables. It is already known that the square wave from block V_{21} is passing through the complete functional group. One more test between blocks V_{40} and V_{41} isolates the faulty circuit.

Divergent Circuit

A divergent circuit is the opposite of a convergent circuit; the output of a single circuit feeds into the inputs of two or more other stages. The illustration of Fig. 8-30 shows that the

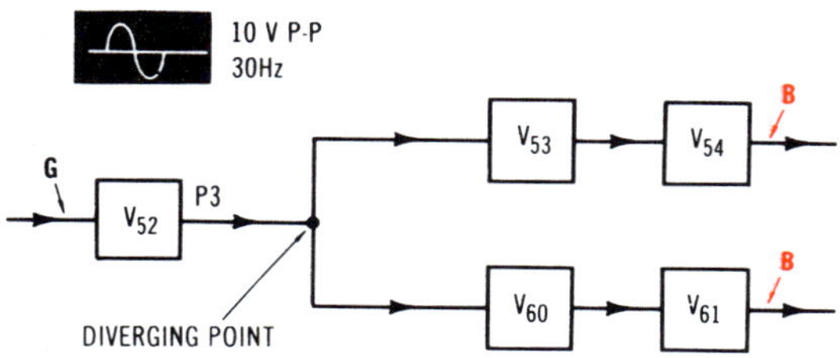

Fig. 8-30. Divergent circuit.

input to the divergent circuit (V_{52}) is good and that the output should be a 30-Hz sine wave at 10 volts peak-to-peak. If the B mark indicates no output at either point, the first test should be made at the output (pin 3) of block V_{52}. If there is no output at this point, enclose V_{52} with a G and a B, and Step 4 is concluded.

Suppose that the divergent circuit conditions were as shown in Fig. 8-31. Actual measured waveforms are shown at the appropriate points as a result of Step-3 testing.

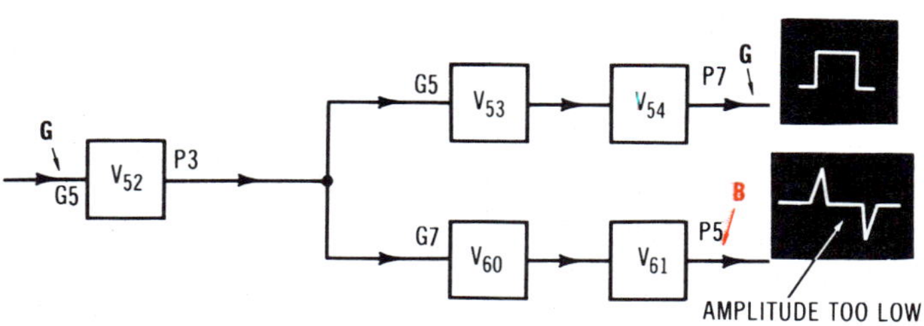

Fig. 8-31. Example circuit conditions.

Q8-13. What are the operating conditions of blocks V_{53} and V_{54}?

Q8-14. Is block V_{52} in a good or bad operating condition?

Q8-15. Based on conditions in the diagram of Fig. 8-31, where should the first Step-4 test be made?

Your Answers Should Be:

A8-13. The output mark of G for that channel indicates that blocks V_{53} and V_{54} are **operating properly.**

A8-14. Since the output of block V_{54} is a good signal, block V_{52} is probably in **good** operating condition.

A8-15. The first Step-4 test should be made at the **output of block V_{60}** or the **input of block V_{61}.**

If the test mentioned in Answer A8-15 is favorable, the G mark is moved up, and V_{61} is enclosed with a G and a B. If the test requires that a B be moved to the output of block V_{60}, a second test at this input will enclose either block V_{60} or V_{52} as the faulty circuit. If it is found that block V_{52} is producing the improper waveform, then the G analysis of the square-wave output from block V_{54} is not very accurate, or the faulty output of V_{52} is sufficient as an input to that channel.

Switching Circuits

In many types of equipment, two or more circuits or channels may be switched individually to another channel. This is shown by the circuit of Fig. 8-32.

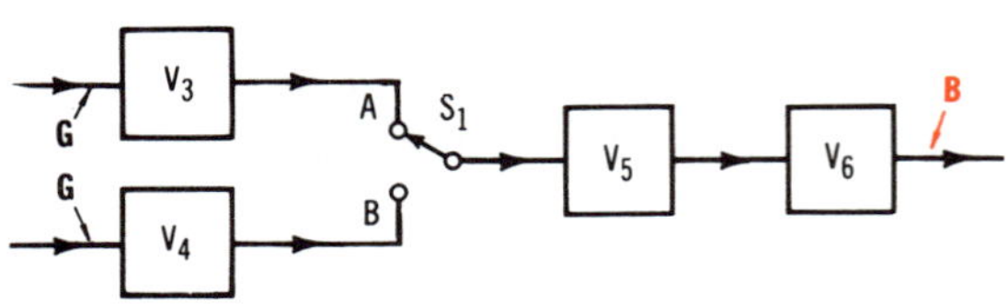

Fig. 8-32. Switching circuit.

With switch S_1 in position A, Step-3 tests reveal good inputs to blocks V_3 and V_4 but a bad output from block V_6. The first test to make in Step 4 is a reading at the output of block V_6 with switch S_1 in position B. If the reading is what it should be, blocks V_5 and V_6 are good and the B mark can be moved to the output of block V_3.

If the output of block V_6 is found to be bad, none of the enclosing marks can be moved, but additional information has been obtained about the probable location of the fault. It is improbable that both block V_3 and V_4 would go bad simultaneously. This conclusion is verified by making the next test at the input of block V_5. If, as suspected, the check is good, then either V_5 and V_6 is faulty, and the obvious enclosing test is made.

The switching-circuit channels can also appear in reverse order, such as a single channel capable of being switched into one of two or more channels.

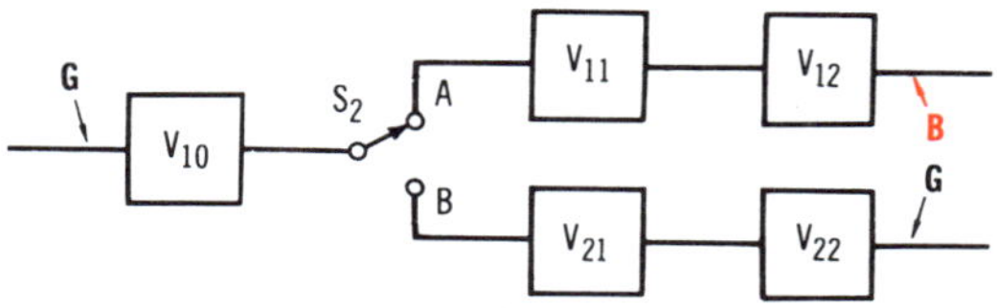

Fig. 8-33. Another switching circuit.

In the diagram of Fig. 8-33, when switch S_2 is switched to position B, the conditions indicate that blocks V_{10}, V_{21}, and V_{22} are good. This leaves blocks V_{11} and V_{12} in the circuit group as the only suspects. The quickest way to determine which is bad is by testing between them. These are the typical circuit combinations that you will encounter while troubleshooting. You have seen that by combining technical knowledge with common sense, the number of enclosing tests needed to isolate a faulty circuit in a functional group can be kept at a minimum.

Enclosing a Tv Receiver Function

Assume that the vertical and horizontal function of a tv receiver has proven to be faulty. Circuit conditions are as shown in Fig. 8-34. Horizontal- and vertical-sync pulses appear at the grid of V_{13}, where they are amplified. The sync separator separates the vertical-sync pulses from the horizontal-sync pulses. The vertical and horizontal channels generate waveforms of the proper frequency, shape, and amplitude to cause the electron beam in the crt to sweep the required horizontal lines on the screen. Oscillations continue whether sync

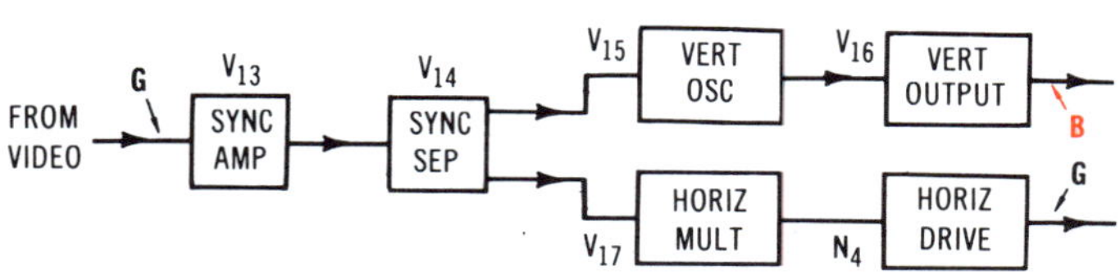

Fig. 8-34. A tv sweep function.

pulses are received or not. This is the reason for the raster appearing on the screen when a station is not tuned in.

The purpose of the sync pulses is to time the start of each line with corresponding events in the tv camera. If this were not done, the picture on the screen would be greatly distorted and unrecognizable.

The symptom noted in Step 1 of the troubleshooting procedure was a thin horizontal line across the face of the screen. Sound was good, and the condition repeated itself on all channels. Steps 2 and 3 earmarked the vertical and horizontal function as being faulty. In the testing process, the input of sync pulses to V_{13} appeared good. So did the output of the horizontal channel. (N_4 is a network of capacitors and resistors that sharpen the shape of the waveform.) There is no measurable waveform at the output of V_{16}. Appropriate enclosure marks are shown on the diagram. Blocks V_{13} and V_{14} are apparently operating properly in accordance with the tests shown. At least the horizontal portion of V_{14} appears to be good. The first test should be made at the input to V_{15}. This allows the **G** or **B** to be moved to that point, depending on test results. From there, only one more test must be made to isolate the single faulty circuit. If the input to V_{15} or the corresponding output from V_{14} is good, the next test should be made at the output of V_{15} or the input of V_{16}.

Step-4 Review

When picking out the faulty stage of a circuit group, symptoms and data from the first three steps are used in making deductions from a study of a servicing block diagram. Enclosure marks employing pencil, weights, or memory are placed at function inputs and outputs to show whether previous tests were good or bad. The enclosure marks are moved in accordance with circuit input or output tests made as the result of a technical and common-sense analysis of the circuit types. The circuit types can be linear, convergent, divergent, or switching.

STEP 5. SEEK OUT AND VERIFY THE CAUSE OF THE TROUBLE

The troubleshooting procedure, thus far, has narrowed the trouble to a single circuit, consisting of a few electronic parts.

The **seek-out** portion of the final step suggests that the faulty part be found and verified as the cause of the trouble.

Analyzing the Output Waveform

The trouble can be narrowed down by analyzing the output waveform of the circuit, making voltage or resistance checks, and/or by substituting a good part for one that is suspected of being bad. Comparing the output waveform actually measured against that of the proper waveform often provides clues as to the location and/or cause of the trouble. The illustration in Fig. 8-35, for example, shows the good and bad outputs of an amplifier. From your knowledge of how an amplifier circuit works, the trouble seems to be in the grid or cathode sections. Thus, examine that portion of the circuit.

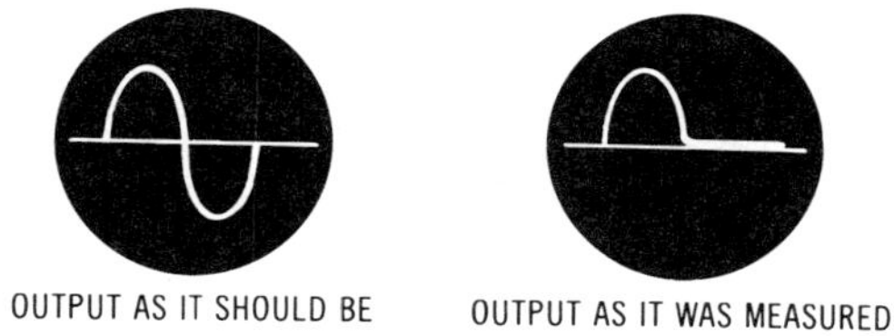

Fig. 8-35. Output waveform.

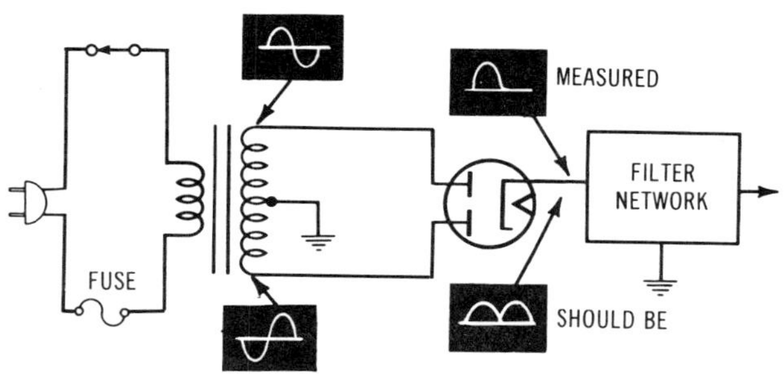

Fig. 8-36. A full-wave rectifier circuit.

Fig. 8-36 shows a faulty full-wave rectifier circuit with measured input and output waveforms. The proper output waveform is also included. Half of the output cycle is missing. A study of the schematic diagram and the waveforms reveal that

Q8-16. What does Step 4 enable you to do?

Q8-17. What is the purpose of Step 5?

Your Answers Should Be:

A8-16. Step 4 enables you to **determine in which circuit** of the equipment **the malfunction is located.**

A8-17. Using Step 5, you can find the **component** that is causing the trouble in the suspected circuit.

you should concentrate your search in the lower plate section of the diode. This type of waveform analysis, in most cases, will help you limit your search to a small area of a circuit. Of equal or even greater importance, a knowledge of the nature of the distortion in an improper waveform will assist you in verifying that the located fault is the actual cause of the trouble.

Making Voltage and Resistance Checks

If an analysis of the output waveform provides a probable location, or if there is no waveform to be analyzed, the next procedure involves making voltage and resistance checks. Some schematics provide both voltage and resistance readings in chart form. If the measured values of a suspected part are not reasonably close to those indicated in the diagram or chart, you have narrowed down the trouble.

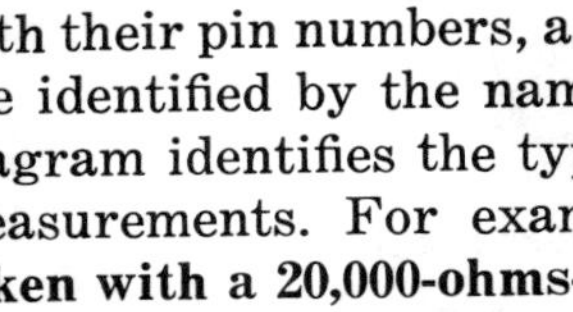

Fig. 8-37. Tube and transistor elements.

In nearly all examples, the elements of a tube are marked with their pin numbers, as shown in Fig. 8-37. Transistor leads are identified by the name of the element. A notation on the diagram identifies the type of instrument used in making the measurements. For example, **dc voltage measurements are taken with a 20,000-ohms-per-volt meter, and ac voltages with a 1000-ohms-per-volt meter.** The schematic diagram may also include the following notations:

1. **Pin numbers are counted in a clockwise direction when viewed from the bottom of the tube socket.**
2. **Measured values are from pin socket to ground unless otherwise indicated.**

3. Transistor terminals should be identified from transistor handbooks or specification sheets. However, these are sometimes also listed on the schematics.
4. Controls are set for normal operation.
5. Component values are given in ohms and microfarads, unless otherwise stated.

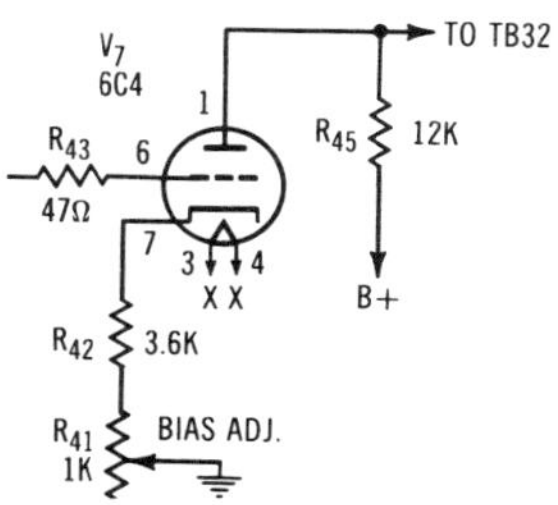

ELEMENT	PIN NO.	VOLTAGE	RESISTANCE
PLATE	1	+250 V	130K
FILAMENT	3	3.15 V AC	0
FILAMENT	4	3.15 V AC	0
GRID	6	+72 V	80K
CATHODE	7	+76 V	5K

Fig. 8-38. Voltage and resistance measurements.

It is generally better to take voltage readings first and resistance readings second, if they are required. If voltages are to be read at all pin numbers, it is best to take them in a sequence of voltage values rather than pin numbers. For example, start with a high scale on the voltmeter and measure the pin having the highest voltage (Fig. 8-38). In this case, it would be the plate, pin 1. Use a scale that will permit a higher reading than the 250 V shown; this will prevent damage to the meter, in case the actual voltage is much higher. Next, measure pins 6 and 7, since they are supposed to have voltage values relatively close together. Finally, pins 3 and 4 can be measured on an ac voltage range.

If you suspected some sections of the circuit from an analysis of the improper output waveform, take voltage readings at those sections first. For example, if your deduction indicated that the waveform was being distorted by an improper grid-to-cathode bias, check those pin voltages first. If the pin 6 measurement is the proper 72 V but pin 7 is 80 V, your deduction would be confirmed and the trouble fairly well isolated.

Q8-18. How would you determine which components are included in pin-to-ground resistance readings?

Q8-19. If you measured zero volts at pin 3 in the diagram of Fig. 8-38, what would be your conclusion?

Your Answers Should Be:

A8-18. All the parts included in the resistance being measured from a pin to ground can be determined by **tracing them out on the schematic diagram.** Make sure you find all the parallel paths.

A8-19. If zero volts were measured on the heater pin, the conclusion would be that **voltage is not available at this pin.**

A LOGICAL APPROACH TO TROUBLESHOOTING

Logical troubleshooting is a step-by-step approach to isolation of a trouble. Television receiver manufacturers sometimes provide helpful information of this type in service manuals. For example, for troubleshooting the picture-tube circuits shown in Fig. 8-39, the manufacturer provides a sequenced procedure. This is given in Fig. 8-40. It is assumed that you've already checked the crt and found it to be good, and that the filaments are lit. The following is an explanation of the procedure given in Fig. 8-40.

1. With a HV probe, check the second anode. If the high voltage is considerably less than 26 kV, the high voltage circuits are at fault. If the high voltage is all right, go on to the next step.
2. Check the high voltage at pins 1 and 3 for 7 kV and 12 kV, respectively. If one or both values are missing, it means a focus-circuit problem. If both are all right, go on to step 3.
3. Check pin 11 for 750 volts at G2. If 750 volts are present, circuit is functioning properly—go to next step. If 750 volts are not present, the G2 circuit needs work.
4. Check pin 9 for +40 volts at G1. If voltage is missing or incorrect, check the G1 supply. If +40 V is present, go to next step.
5. Check pins 10, 12, and 13 for +190 volts, and also for green, red, and blue waveforms (video). If one or more of the 190-V potentials is missing, check the circuit involved; if a waveform is missing, check the driver and output circuits of the missing signal.

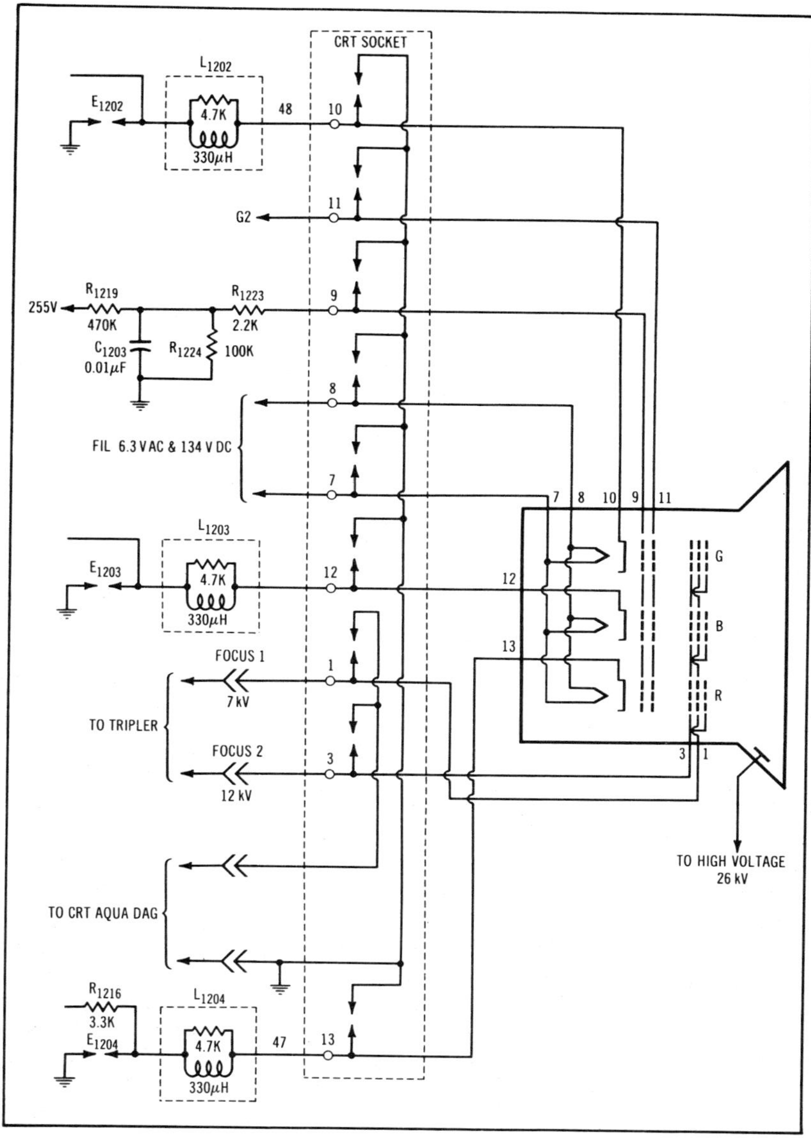

Fig. 8-39. Crt socket and video output circuit.

6. If all the preceding tests check out all right, the crt circuits are functioning properly, so check elsewhere for the problem.

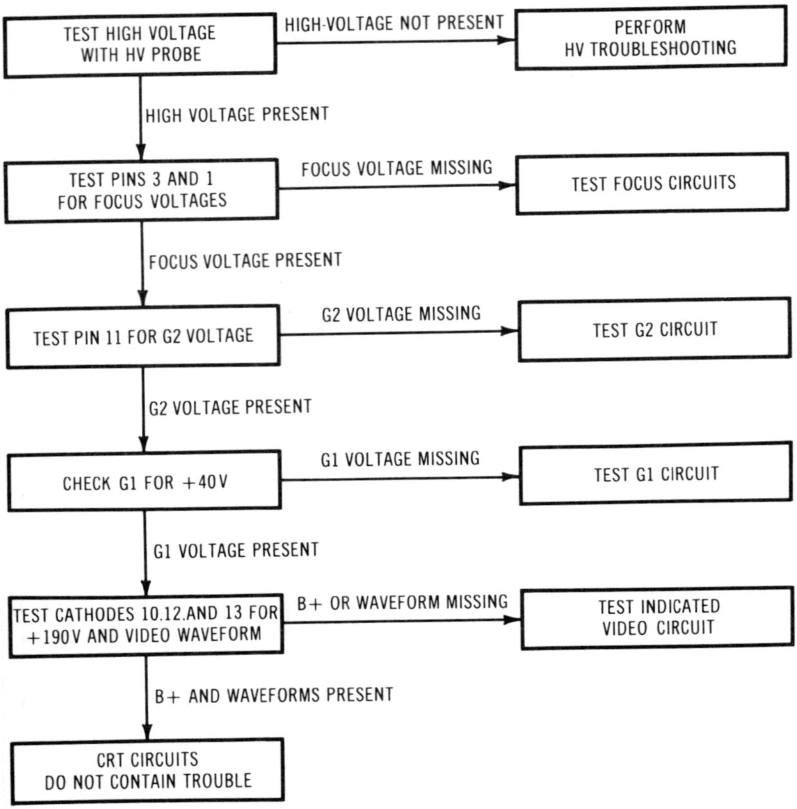

Fig. 8-40. Troubleshooting procedure for crt circuit.

RESISTANCE MEASUREMENTS

Fig. 8-41 is an example of a resistance table as provided by the manufacturer of electronic equipment for servicing that equipment. As stated at the top of the table, the listed measurements were taken with a meter having 0.08 volt maximum between the probe tips. It is important to list this voltage so that the user of the dmm will know whether to use the **Hi Power Ohms** or the **Lo Power Ohms** function for the resistance measurements. Usually, any designation below 0.1 volt means that the Lo Power Ohms range should be used.

However, for checking diode front-to-back ratios and tran-

MEASUREMENTS BELOW TAKEN WITH METER HAVING .08V MAX BETWEEN PROBE TIPS														
ITEM	PIN 1	PIN 2	PIN 3	PIN 4	PIN 5	PIN 6	PIN 7	PIN 8	PIN 9	PIN 10	PIN 11	PIN 12	PIN 13	PIN 14
V101	35 M	NC	27 K	NC	NC	FIL	FIL	27 K	33 K	4 M	NC	27 K	NC	
IC400	33 K	3000	8200	45 K	0	1600	INF	INF	2000	2000	2000	360	9000	360
													PIN 15 40 K	PIN 16 33 K
IC401	9000	900	900	1550	0	5700	1200	1200	INF	750	37 K	37 K	700 K	700 K
													PIN 15 33 K	PIN 16 33 K
IC500	INF	INF	0	0	700	12 K	8500	INF	5400	5400	INF	11 K	5700	50 K
IC1000	INF ≠	360	0											
IC1001	51	750	43 K											

ITEM	E	B	C		ITEM	E	B	C		ITEM	E	B	C	
Q100	39	12 K	60 K /		Q304	95 K	90 K	520		Q703	1000	1800	0	
Q101	INF	INF	INF ≠		Q305	95 K	2200	5600		Q704	1500	3000	23 K /	
Q102	INF	INF	INF ≠		Q400	360	2500	1650		Q705	1500	3000	23 K /	
Q103	.2	1.5	45 K /		Q401	0	25 K	3600		Q706	1500	3000	23 K	
Q105	.68	1500	INF		Q402	0	15 K	6200		Q707	680	200	1500	
Q106	INF	INF	INF		Q500	470	9000	770		Q708	680	50	1500	
Q200	390	90 K	1500		Q600	INF	INF ≠	40 K		Q709	680	40	1500	
Q201	1000	4000	1600		Q601	0	54 K	INF		Q710	0	10 K	INF ≠	
Q202	300	900	770		Q602	5600	INF	INF		Q711	0	15 K /	INF ≠	
Q203	680	4000	850		Q603	0	5600	INF		Q800	360	30 K	50 K /	
Q204	330	680	1600		Q604	INF	INF	INF		Q801	470	3900	53 K /	

Fig. 8-41. Resistance measurements. (*Courtesy Sencore, Inc.*)

sistor junctions, the Hi Power Ohms function is used. Then, the dmm will not show zero ohms for the forward-bias condition as does an analog meter. Instead, the dmm will indicate the voltage drop across the forward-biased junction. The voltage drop will be 0.3 volt for germanium transistors and 0.6 volt for silicon transistors.

Referring back to Fig. 8-41, note that resistance readings are listed for various pins and electrodes of several devices. There are resistance readings for one vacuum tube (V101), five integrated circuits (IC400, IC401, etc.), and several transistors (Q100, Q101, etc.). The pin numbers indicate measurements from each pin to ground for the vacuum tube and the ICs. For example, for IC400, pin 1 to ground should measure 33K ohms, pin 2 should measure 3000 ohms, pin 3 should measure 8200 ohms, and so on. Next, look at transistor Q703. The emitter-to-ground reading should be 1000 ohms, the base should read 1800 ohms, and the collector should measure 0 ohms to ground.

Sometimes, voltage values are also given in service literature. The data usually is in the form of a table, such as the following "Voltage Chart." Note that the information alongside the table states that the voltages are negative and are measured with respect to pin 6 on the power plug. Further, the measurements were taken with a vtvm having an 11-megohm input resistance.

Voltage Chart E	B	C	Notes
6.2	5.6	6.0	All voltages measured with respect to pin 6 on power plug and are negative.
7.8	7.2	5.6	Voltages measured with a vtvm having 11-megohm input resistance.
8.9	9.5	12.2	

In another type of presentation of voltage and resistance readings, voltages are placed near the tube elements and the resistance measurements are shown in chart form (Fig. 8-42).

As in an earlier example, part numbers and their values are shown on the schematic. If the trouble is isolated to one section of the circuit, the individual parts can then be measured. If, for example, something other than 25 V and 920 ohms were read from pin 3 to ground, either capacitor C_3 or resistor R_4 would be suspected of causing the trouble.

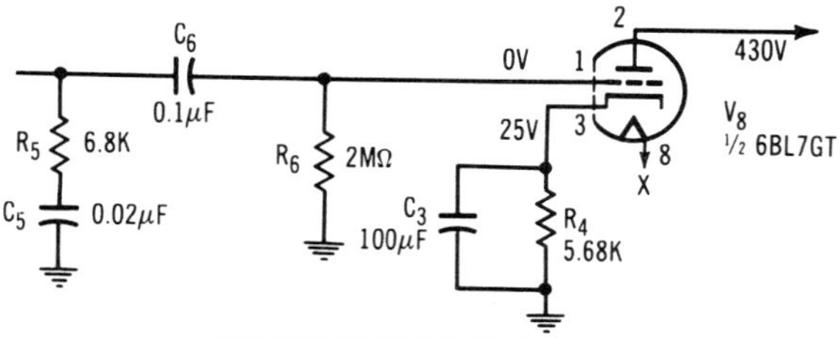

RESISTANCE MEASUREMENTS

STAGE	TUBE	PIN 1	PIN 2	PIN 3	PIN 4	PIN 5	PIN 6	PIN 7	PIN 8
V_8	6BL7GT	2MΩ	3.5K	920Ω	2.4MΩ	2.6MΩ	0	0	0.1Ω

Fig. 8-42. A different way of showing voltage and resistance measurements.

If resistor R_4 is to be measured, one of the capacitor leads must be separated from the resistor. Otherwise, if ohmmeter test leads were placed across the resistor, they would still measure the parallel resistance of the two parts. In this case, a capacitor lead could be unsoldered either from ground or from the tube-socket pin. If resistor R_4 does not measure close to 5.68K, it may be the cause of the trouble.

Q8-20. **A step-by-step approach to the isolation of a problem is called ________ troubleshooting. ing.**

Q8-21. **Before going through the step-by-step procedure previously described, you should first check to make sure that the ___ is good.**

Q8-22. **If you measure plus __ V at pin 9 for G1, it is all right to go on to the next step.**

Q8-23. **In a resistance table, it is important to know both the type of meter used in making the measurements and the maximum voltage between the probe tips. This information is needed so that the user of a dmm will know whether to use the ________ or the ________ function of the dmm.**

Q8-24. **The voltage drop across the forward-biased junction of a germanium transistor should measure _ volt and the voltage drop across the forward-biased junction of a silicon transistor should measure _ volt.**

Your Answers Should Be:

A8-20. A step-by-step approach to isolation of a problem is sometimes called **logical** troubleshooting.

A8-21. Before going through the step-by-step procedure previously described, you should first check to make sure that the **crt** is good.

A8-22. If you measure plus **40** V at pin 9 for G1, it is all right to go on to the next step.

A8-23. In a resistance table, it is important to know both the type of meter used for making the measurements and the maximum voltage between the probe tips. This information is needed so that the user of a dmm will know whether to use the **Lo Power** or the **Hi Power** function of the digital multimeter.

A8-24. The voltage drop across the forward-biased junction of a germanium transistor should measure **0.3** volt and the voltage drop across the forward-biased junction of a silicon transistor should measure **0.6** volt.

Part Substitution

It is often necessary to substitute a known good part for one suspected of being faulty. Capacitor C_3 in the circuit of Fig. 8-42 is an example. Replacing it with a new capacitor of the same value will confirm whether it is good or bad. The same reasoning applies to other parts, including tubes and transistors. As you may recall from the preceding chapter on tube testers, one of the valid methods of checking a faulty tube is to substitute a good one.

Verifying the Cause

Although a faulty part can actually be located by the preceding methods, Step 5 is not completed. The nature of the fault must be compared with and verified by the trouble symptoms that were obtained in the preceding steps. Thus, if an open resistor, a shorted capacitor, a weak tube, etc., adequately explains the improper waveforms and the trouble symptoms, then you can feel reasonably sure that you have found the actual cause of the trouble and you can make the necessary repair(s).

However, if the nature of the trouble does not substantiate the distorted waveforms or other trouble symptoms, you have not found the faulty component or, in some cases, you have found only one of the faulty components. For example, a slight change in the value of a plate resistor does not explain the loss or flattening of a half-cycle waveshape in an amplified sine wave waveform.

The faulty component you have isolated may have been the result of a fault in another part of the circuit or even in an adjacent circuit. The narrowing-down procedure may have uncovered a cathode resistor whose measured resistance deviates greatly from its rated value. In addition, if the resistor is badly charred, it is evident that the resistor has been passing an excessive amount of current. The cause could have been a gradual decrease in resistance over a period of time, allowing more and more current to pass until the charred condition resulted. However, the increase in current could also have been caused by a faulty component in another part of the circuit. A decrease in the plate or screen resistance would also cause excessive current to flow in the circuit and damage the cathode resistor.

If, in such a case, the cathode resistor were replaced without verifying the cause of the trouble, the same trouble symptoms would repeat themselves after a period of time. Always verify that the isolated fault explains the trouble symptom(s) fully and that it is the actual cause of the circuit malfunction.

Repair

You may have noted that the word **repair** was not included in the troubleshooting steps. Replacing a part, resetting an adjustment, or restoring a connection is actually not a part of troubleshooting. Troubleshooting includes all the processes required to isolate the faulty condition. Once the trouble has been found and verified, then the repair to the circuit can be made.

WHAT YOU HAVE LEARNED

1. Troubleshooting is the process of locating a fault in a piece of equipment.
2. To become a good troubleshooter, you must:
   (A) Know enough about electronic principles to use them in determining how equipment operates.
   (B) Know enough about the use of test equipment to make and interpret test readings properly.
   (C) Know enough about electronics to extract desired information from a technical manual or service folder.
   (D) Know enough about the logical troubleshooting procedure, **STEPS,** to apply it well.
3. A good method of troubleshooting is a systematic orderly process called logical troubleshooting.
4. The logical troubleshooting procedure consists of five parts. The initial letters of these parts spell the word **STEPS.**

   STEP 1. Search for all trouble symptoms.
   STEP 2. Trace out all probable faulty functions.
   STEP 3. Expose the single faulty function.
   STEP 4. Pick out the faulty circuit.
   STEP 5. Seek out and verify the cause of the trouble.

Index